NANOWIRES

Complete Handbook

NANOWIRES
Complete Handbook

by
Maya Sheikh | Zahra Ahamad

Wordpen Academics
Publishing Science That Matter

ISBN: 978-93-6084-872-9

Published by : **WORDPEN ACADEMICS**
4736/23, Ansari Road, Darya Ganj,
New Delhi – 110 002
Phone:
E-mail:

Printed at **:** **Replika Press Pvt. Ltd.**

Exclusive Distributor: ***Bio-Green Books, New Delhi***

Preface

The "Nanowires: Complete Handbook" stands as a comprehensive compendium, delving into the intricate world of nanowires, their synthesis, properties, and applications. Nanowires, with their unique physical and chemical properties at the nanoscale, represent a cornerstone of nanoscience and nanotechnology, offering a wealth of opportunities for scientific exploration and technological innovation.

In recent years, nanowires have emerged as key building blocks for a wide range of applications, spanning electronics, photonics, energy conversion, sensing, and biomedicine. Their high aspect ratio, tunable properties, and compatibility with existing fabrication techniques make nanowires a versatile platform for engineering novel devices and systems with enhanced performance and functionality.

As editors of this volume, we have endeavored to compile a diverse array of perspectives, methodologies, and applications from leading experts and researchers in the field of nanowires. Through a blend of theoretical insights, experimental techniques, and practical applications, this handbook covers a wide range of topics, including nanowire synthesis methods, characterization techniques, device fabrication, and emerging applications in various fields.

Our overarching goal in presenting this handbook is to provide readers with a comprehensive and up-to-date resource that serves as a guide for researchers, engineers, educators, and students interested in exploring the multifaceted world of nanowires. By showcasing the latest developments and trends in nanowire research, we aim to inspire curiosity, foster collaboration, and accelerate innovation in this exciting field.

We extend our sincere gratitude to all the contributors who have generously shared their expertise, experiences, and insights in the field of nanowires. It is our hope that this handbook will serve as a valuable reference and guide for researchers and professionals alike, as they navigate the frontiers of nanoscience and nanotechnology and contribute to the advancement of knowledge and technology in this rapidly evolving field.

Maya Sheikh | Zahra Ahamad

Contents

Preface v

1. Nanowires of Fe/Mgo/Fe Encapsulated in Carbon Nanotubes 1

1. Introduction—2. Experiment—3. Structural characterization—4. Results from magnetic measurements—5. Discussion—6. Conclusions—References

2. Diamond Nanowire Synthesis, Properties and Applications 15

1. Introduction— 2. Synthetic strategies for DNW growth—3. Structures and properties of DNWs—4. Applications of DNWs—5. Conclusions and perspectives— References

3. Analysis of Electrochemical and Structurally Enhanced Limn2O4 Nanowire Cathode System 37

1. Introduction— 2. Experimental procedure—3. Characterization and analysis — 4. Results and discussions—5. Conclusion—References

4. ZnO Nanowires Growth Direction and Parameters Affecting their Surface Morphology 49

1. Introduction—2. ZnO nanowire synthesis—3. Hydrothermal method—4. ZnO nanowire synthesis mechanism—5. Parameters affecting ZnO nanowire growth and morphology—6. Conclusion—References

5. Nanowires for Room-Temperature Mid-Infrared Emission 71

1. Introduction—2. The growth—3. Structural analysis—4. 4-K micro-PL measurements—5. PL power dependence—6. Development of a band structure for the MQWs—7. PL temperature dependence—8. Conclusion—References

6. Parasitic Capacitances on Scaling Lateral Nanowire **91**

1. Introduction—2. GAA-NW device—3. Simulation methodology—4. Parasitic capacitance estimation—5. Ring oscillator simulation—6. Conclusion—References

7. Magnetoresistance and Structural Characterization of Electrospun La¹–Xsrxmno³ Nanowire Networks **109**

1. Introduction—2. Experimental procedures—3. Results and discussion—4. Conclusion—References

Index

1

Nanowires of Fe/Mgo/Fe Encapsulated in Carbon Nanotubes

1. Introduction

In TMR, electrical current flows across a barrier of nanometric thin insulator layer between two ferromagnetic metal electrodes when an external magnetic field is applied parallel to the trilayer surface. TMR is one of the few examples of macroscopic quantum mechanical phenomena that have no classical explanation. TMR has a wide array of applications including magnetic random access memory for futuristic quantum computer and ultrasensitive sensors [1]. TMR is the basic building block of magnetic tunnel junctions (MTJs); specifically Fe/MgO/Fe TMR having one of the highest magnetoresistance (MR) ratio is of current interest as evidenced by several studies of its magnetic properties [2, 3, 4, 5, 6, 7]. Theory predicts several thousand percent MR ratio for Fe/MgO/Fe TMR trilayers when an interface is modeled to possess an abrupt change from Fe to MgO without any diffusion at the interface [3]. When diffusion at the interface is introduced in the theoretical modeling by way of interface oxidation, the MR ratio dropped to 1000% [2]. First-principle modeling showed that diffusion at the interface plays a major role in decreasing the calculated value of the MR ratio to the extent when only 16% of Fe is replaced by Mg and vice versa; the MR ratio agreed with experimental values for higher values of the insulator (MgO) thickness [4]. The noncollinear nature of the atomic moments with the bulk magnetization at the interface of Fe and MgO will perhaps affect the net anisotropy, or it may be a spin scattering locality [5]. Experimental measurements showed that single-crystal Fe/MgO/Fe has an MR ratio of 180% at room temperature and 247% at 20 K [6]. It was also shown experimentally that for crystalline MgO (001), the barrier between the electrodes resulted in a 220% MR ratio at room temperature [7]. The symmetry of electronic wave functions plays a paramount role in coherent spin-polarized tunneling, which gives rise to enormous TMR effects. Shape anisotropy plays an important role in coherent spin-polarized tunneling of electrons across the insulator barrier; thus, switching the geometry of TMR from planar films to nanowires will increase the coherence thus increasing the MR ratio [8, 9]. The importance of shape anisotropy in ferromagnet/insulator/ferromagnet trilayer structures in particular for Fe/MgO/Fe and the effect of encapsulation of a TMR by carbon tubes are discussed in this chapter based mainly on magnetic measurement results and subsequent interpretations.

TMR-based magnetic tunnel junctions are theoretically expected to exhibit an extremely high MR ratio due to coherent tunneling [2]. When the coherency of electron wave functions is conserved during tunneling, only conduction electrons whose wave functions are totally symmetrical with respect to the barrier-normal axis are connected to the electronic states in the barrier region and have significant tunneling probability. The MR ratio is defined as follows:

$$\text{MR ratio} = \left(R_{ap} - R_p\right)/R_p,$$

where R_p and R_{ap} are the tunnel resistance when the magnetizations of the two electrodes are aligned in parallel and antiparallel, respectively (Figure 1).

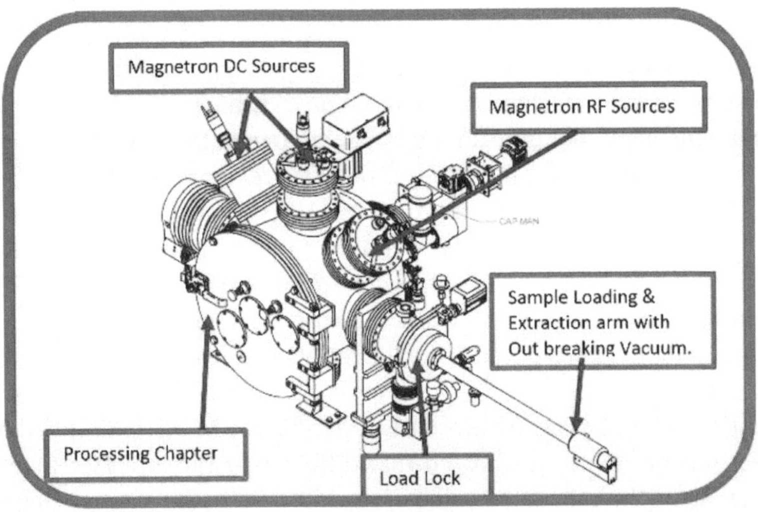

Figure 1. Magnetron sputtering tool with 3 DC/1 RF sources and a load-lock.

A TMR system reveals unique properties with attractive effects for technological applications. Besides giant and tunneling magnetoresistance, it presents other remarkable effects such as antiferromagnetic exchange coupling, oscillatory behavior of exchange coupling and biquadratic exchange couplings. Nanometric ultrathin films exhibit an out-of-plane uniaxial surface anisotropy sufficient to overcome the demagnetizing field. This feature makes it important for high-density magnetic media, highly sensitive sensors and parts for a quantum computer. This system in addition to its enormous potential for technological applications is an attractive research object in nanomagnetism.

Samples were prepared in the form of nanometric thin films and nanocolumns as shown in Figure 2. Nanocolumn arrays were fabricated with different shapes and in-plane orientations by glancing angle deposition (GLAD). This will provide shape anisotropy, which will compete with surface and volume anisotropy. Many unique and fascinating properties have already been demonstrated by nanocolumns synthesized using GLAD, such as superior mechanical toughness, higher luminescence efficiency, enhancement of thermoelectric figure of merit and lowered lasing threshold. Homogeneous nanowires and nanowire networks have been previously used as chemical sensors, field-effect transistors and inverters, photodetectors, light-emitting diodes, lasers and logic gates. Very recently, by altering the compositions of the nanostructures during fabrication, super-lattice nanowire has been demonstrated, which can greatly increase the versatility and application of these building blocks in nanoscale electronic, photonic, and biological applications. Fe nanocolumns were used to synthesize metal-assisted protein crystallization [10]. Possible applications, including thermoelectrics, nanobarcodes, injection lasers and one-dimensional waveguides, could be implemented through these super-lattice nanostructure building blocks. One very important issue associated with these studies is how to assemble one-dimensional nanostructures in an effective and controllable way. GLAD produces columnar structures through the effect of shadowing during film growth, while the substrate rotation controls the shape of the columns.

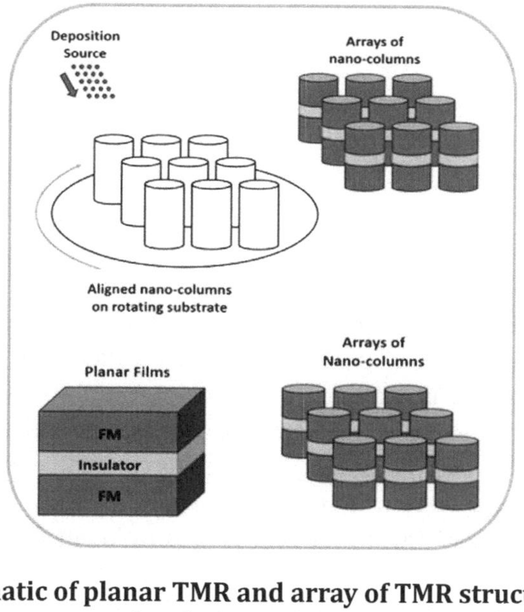

Figure 2. Schematic of planar TMR and array of TMR structures. Both sets of samples were grown using the same magnetron sputtering deposition tool. The planar films will establish initial growth conditions for arrays of nanocolumns.

2. Experiment

Nanowires, planar nanometric thin films and nanocolumns of Fe/MgO/Fe were synthesized using magnetron DC/RF sputtering, AXXIS sputtering tool from Kurt J. Lesker Company. Fe target of purity 99.9% and MgO of purity 99.95% were sputtered using high-purity Ar. Nanowires were synthesized by magnetron DC and RF sputtering in the interior cylindrical space of carbon nanotubes by magnetron DC and RF sputtering.

Thermal chemical vapor deposition method (TCVD) was used to grow vertically aligned CNTs on the SiO2 substrate thus filled with Fe/MgO/Fe in this experiment [11]. TCVD involves exposing silica structures to a mixture of ferrocene and xylene at 770°C for 600 s. In TCVD, the furnace is pumped down to about 200 mTorr in an argon bleed, and then, the temperature is raised to 770°C. The solution of ferrocene dissolved in xylene of concentration about 0.01 g/ml and was preheated in a bubbler to 175°C and then passed through the tube furnace. The furnace is cooled down to room temperature, and tips of the CNTs were exposed to plasma to open the nanotubes for the next procedure. The open-ended CNT tips were filled with Fe/MgO/Fe using DC magnetron for the Fe and RF sputtering method for the insulator (MgO) at a substrate temperature of 100°C. This substrate temperature, 100°C, yielded in planar samples the highest value of coercive field compared to several other synthesis substrate temperatures [12].

Nanowires were synthesized by magnetron DC and RF sputtering in the nanometric interior cylindrical volume of carbon nanotubes at a substrate temperature of 100°C.

Nanometric thin films were epitaxially grown on the MgO (100) substrate of dimensions 5 mm × 5 mm × 0.5 mm using magnetron DC and RF sputtering at several temperatures. All substrates were degassed at 350°C in vacuum of 0.1 μTorr for 1800 s, and samples were pre- and post-annealed at a preselected deposition temperature for 1800 s in vacuum. The source substrate distance was kept fixed at 30 cm, and the substrate surface normal was kept at 45° with a line connecting the center of the sample to the center of the target, while being rotated at a constant rate of 20 rpm for uniform deposition. Under these conditions, epitaxial Fe grows on MgO (100) due to a good lattice match of MgO and Fe, and weak interface interaction [13, 14] free standing Fe is formed. The deposition rate for Fe was 0.17 nm/s as calibrated by the deposition time versus thickness measurements for Fe films several hundred nm thick. In my previous research on thin films of Fe/MgO/Fe, several planar samples were synthesized at several substrate temperatures [12]. The film synthesized at 100°C has the highest saturation magnetization.

Nanocolumns of Fe/MgO/Fe were synthesized at a glancing angle of 70°, and other parameters remained the same as in the synthesis of thin film (Figure 3).

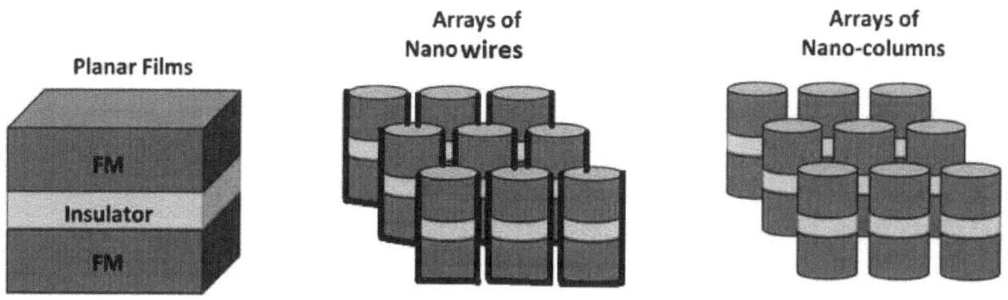

Figure 3. Schematic of planar TMR and array of TMR structure.

All sets of samples were grown using the same magnetron sputtering deposition tool. The planar films established initial growth conditions for the arrays of nanowires and nanocolumns.

3. Structural characterization

Thin film samples of Fe/MgO/Fe on MgO (100) substrates were characterized by XRD (miniflex Rigaku X-ray diffraction of 40 kV/40 mA) using CuKα radiation in θ–2θ geometry (Figure 4).

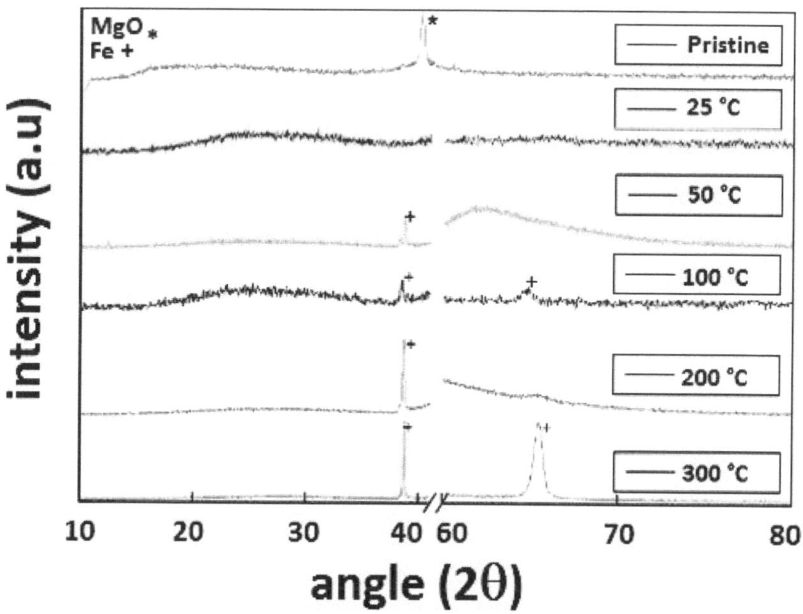

Figure 4. XRD of Fe/MgO/Fe/MgO (100) synthesized at several substrate temperatures and of pristine MgO (100).

Surface morphologies of nanowires of Fe/MgO/Fe grown in the interior cylindrical space of CNTs were characterized by SEM/STEM in a previous study [12]. The SEM/STEM scan depicts a uniform composition.

X-ray absorption spectroscopy (XAS) measurements on thin films and nanowires of Fe/MgO/Fe carried out at beamline 4UB at the National Synchrotron Light

Source (NSLS) in Brookhaven National Laboratory (BNL) by the author revealed the existence of a missing shoulder at the Fe L3 edge in the nanowire spectrum at 717 eV. At the Fe L2 edge, the double peaks at 723.5 eV are of equal height for the nanowires and a leading peak in the film's spectra.

4. Results from magnetic measurements

In a previous work, it was shown that magnetic force microscopy (MFM) of nanowires of Fe/MgO/Fe revealed the presence of stripe domains [12]; stripe domain was totally absent in films, and it was slightly present in nanocolumns. This shows that the moments are completely aligned in plane in case of thin films. In nanowires, the magnetic moment has a significant out-of-plane component.

X-ray magnetic circular dichroism (XMCD) measurements of nanowires of Fe/MgO/Fe were carried out at beamline 4UB at NSLS in BNL [12]. Background-corrected XMCD signal shows that the nanowires' XMCD signal at the Fe L3 and L2 edges is larger [12]. Moreover, a switching between the two edges occurs, and at around 712 eV of photon energy, the intensity of the nanowires' XMCD is smaller [12].

Vibrating sample magnetometer (VSM) measurements were carried out using Vector Magnetometer Model 10 VSM system from MicroSense equipped with 3 T electromagnet (Figures 5–7). In these figures, the coercive field, saturation and remanent magnetization values of thin films, nanowires and nanocolumns of Fe/MgO/Fe synthesized at the substrate temperature of 100°C versus the angle between the applied magnetic field and the surface normal are shown.

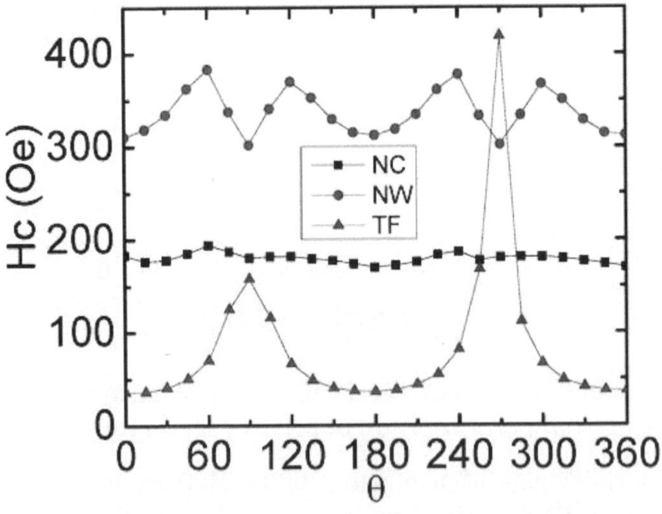

Figure 5. Coercive field (Hc) of nanocolumns (NC), nanowires (NW) and thin films (TF) of Fe/MgO/Fe at several angles between applied field and the sample surface varied between 0 and 360°.

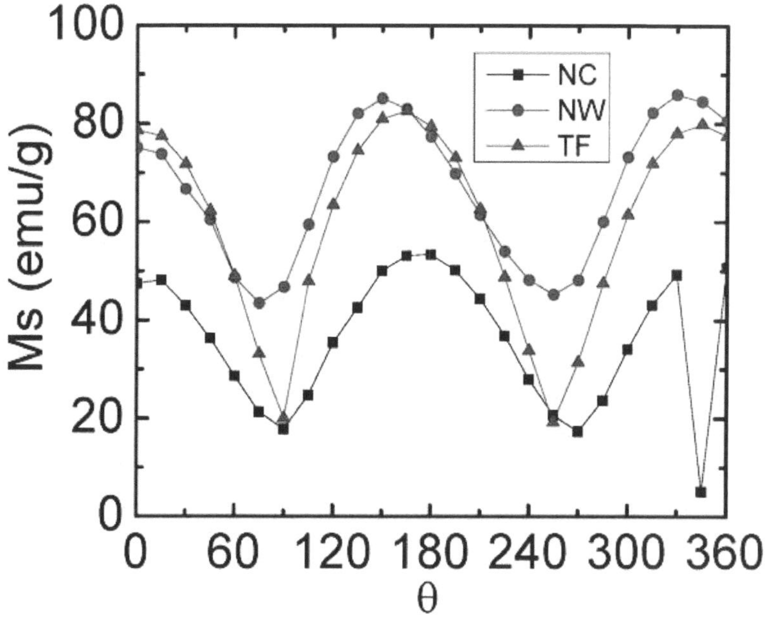

Figure 6. Saturation magnetization (Ms) of nanocolumns (NC), nanowires (NW) and thin films (TF) of Fe/MgO/Fe at several angles between applied field and the sample surface varied between 0 and 360°.

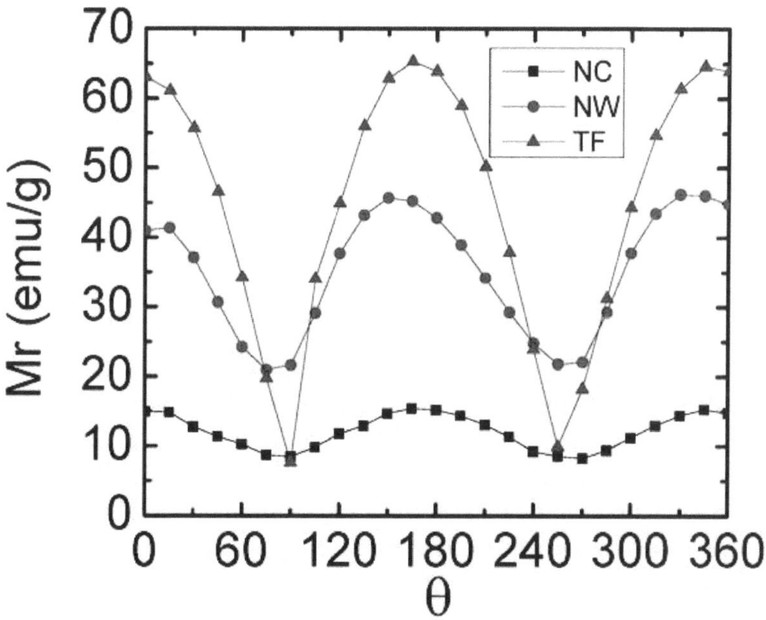

Figure 7. Remanant magnetization (Mr) of nanocolumns (NC), nanowires (NW) and thin films (TF) of Fe/MgO/Fe at several angles between applied field and the sample surface varied between 0 and 360°.

In Figure 5, the coercive field (Hc) of nanocolumns (NC), nanowires (NW) and thin films (TF) of Fe/MgO/Fe at several angles between the applied field and the sample surface varying between 0 and 360° is depicted. The coercive field value of nanowires is higher than both nanocolumns and thin films. At an angle of 270°, there is an outlier value for thin films that is higher than both nanowires and nanocolumns value of Hc at 270°. The Hc values of nanowires oscillate with a period of 90°, whereas those of thin films oscillate with a period of 180°. For nanocolumns, the coercive field values are constant at all angles. The in-plane value of the coercive field of nanowires is higher by 754% than planar thin films.

Saturation magnetization (Ms) of nanocolumns (NC), nanowires (NW) and thin films (TF) of Fe/MgO/Fe at several angles between the applied field and the sample surface varying between 0 and 360° is depicted in Figure 6. The saturation magnetization of all three oscillates with a period of 180°. The saturation magnetization of nanowires is slightly higher than nanocolumns and thin films at all angles.

Remanent magnetization (Mr) of nanocolumns (NC), nanowires (NW) and thin films (TF) of Fe/MgO/Fe at several angles between the applied field and the sample surface varying between 0 and 360° is shown in Figure 7. The remanent magnetization of all three oscillates with a period of 180°. The remanent magnetization of thin films is higher than nanowires and nanocolumns at all angles except at 90 and 270°.

Magnetic torque measurements for nanocolumns of Fe/MgO/Fe synthesized at 100°C for several applied fields were carried out using the EV7 torque magnetometer (TMM) system equipped with a 2 T electromagnet (Figure 8). Torque magnetometer measurements of thin films and nanowires results from the previous study show a similar trend such as a twofold symmetry [12]. TMM measurements of nanocolumns of Fe/MgO/Fe synthesized at 100°C for several applied fields are shown in Figure 8(a) and in the polar plot in Figure 8(b). TMM measurements for planar film and nanowires of Fe/MgO/Fe both synthesized at 100°C were shown to have pronounced twofold symmetry in the rotational hysteresis loop that appeared for an applied field of 5000 Oe and higher [12]. Here, it is shown for the first time that nanocolumns also show pronounced twofold symmetry in the rotational hysteresis loop for an applied field of 5000 Oe and higher (Figure 8). Torque on the film is 20 times higher than on nanowires for the same applied field [12]. Figure 8 shows that TMM measurements for nanocolumns in the polar plot along equivalent crystallographic direction yields torque on the nanocolumns are 10 times higher than on nanowires for the same applied field.

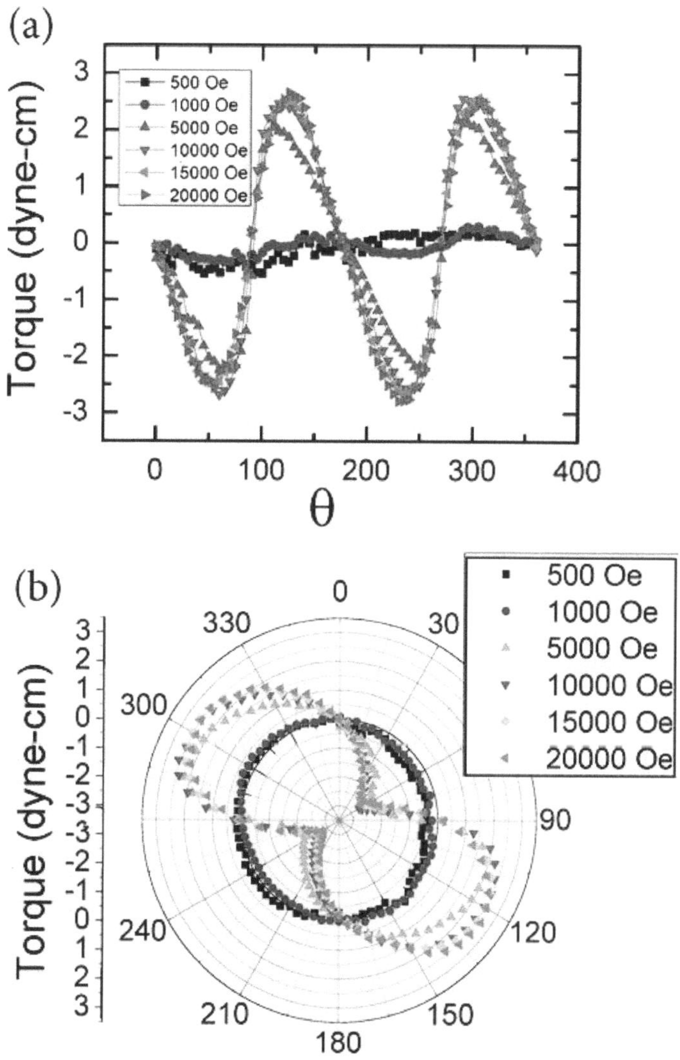

Figure 8. (a) TMM measurements of nanocolumns of Fe/MgO/Fe at several fields while the angle between applied field and the sample surface rotate between 0 and 360°. (b) The same figure in part a in polar plot.

Figure 9 depicts Mr/Ms ratio measured using VSM at several angles with respect to an applied field for thin films, nanowires and nanocolumns of Fe/MgO/Fe synthesized at 100°C. External magnetic field applied along equivalent crystallographic directions did not produce equivalent hysteresis loops. This is obvious when comparing Mr/Ms ratio obtained for symmetric orientations, every 15° from 0 to 360° as shown in Figure 9. Mr/Ms ratio extracted from VSM measurements at every 15° between 0 and 360° for planar film, nanowires and nanocolumns of Fe/MgO/Fe synthesized at 100°C is shown in Figure 9. In all three cases, a twofold symmetry is observed.

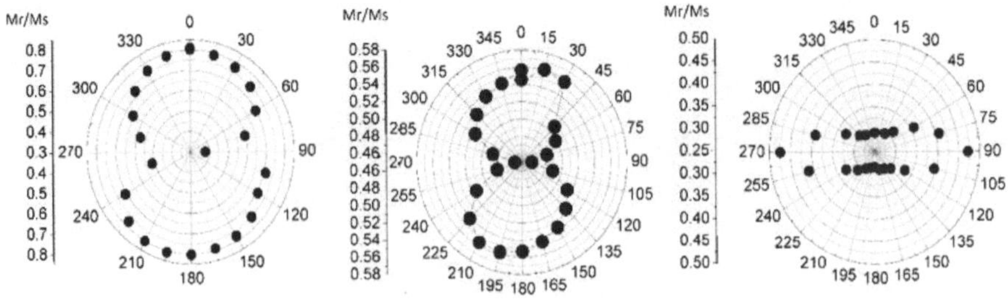

Figure 9. (a) Mr/Ms measurements of thin films Fe/MgO/Fe at several fields. (b) Mr/Ms measurements of nanowires of Fe/MgO/Fe at several fields. (c) Mr/Ms measurements of nanocolumns of Fe/MgO/Fe at several fields.

Figure 10 depicts a coercive field measured using VSM at several angles with respect to an applied field for thin films, nanowires and nanocolumns of Fe/MgO/Fe synthesized at 100°C. Figure 10 depicts a coercive field measured using VSM at several angles with respect to an applied field for thin films, nanowires and nanocolumns of Fe/MgO/Fe synthesized at 100°C. A strong dependence on an angle was observed in the films than in the nanowires. Coercive field (Hc) value extracted from VSM measurements at every 15° between 0 and 360° for planar films, nanowires and nanocolumns of Fe/MgO/Fe synthesized at 100°C is shown in Figure 10. In all three cases, different symmetries are observed. Nanowires depict a fourfold symmetry, whereas a well-defined twofold symmetry is observed in nanocolumns. The conspicuous symmetry difference in nanowires could be due to shape anisotropy and also due to the hybridization that occurs between the π-electronic states of carbon and 3d bands of the Fe surface.

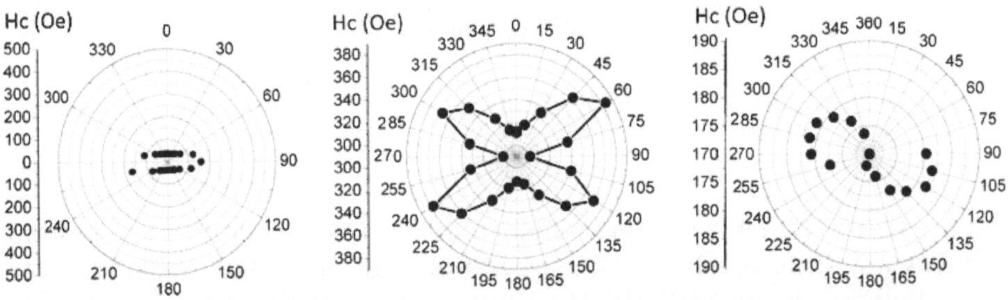

Figure 10. (a) Hc measurements of thin films of Fe/MgO/Fe at several fields. (b) Hc measurements of nanowires of Fe/MgO/Fe at several fields. (c) Hc measurements of nanocolumns of Fe/MgO/Fe at several fields.

Figure 11 depicts temperature-dependent magnetization measurements of planar nanometric thin films of Fe/MgO/Fe/MgO (100) synthesized at 100°C. The planar thin film with the highest value of Hc as shown in Figure 11(a) is the sample synthesized at 100°C. Low-temperature VSM measurements of thin films of Fe/MgO/Fe synthesized at 100°C are shown in Figure 11; as expected, all three magnetization parameters Hc, Mr, and Ms increased as temperature is lowered; more pronounced increase was observed in Hc. As shown in Figure 11(a), the value of Hc increased by 61%, Mr increased by 16% and Ms increased by 11% at 80 K compared to their room temperature values, respectively. Figure 11(b) depicts hysteresis loops of the Fe/MgO/Fe planar film at low temperatures.

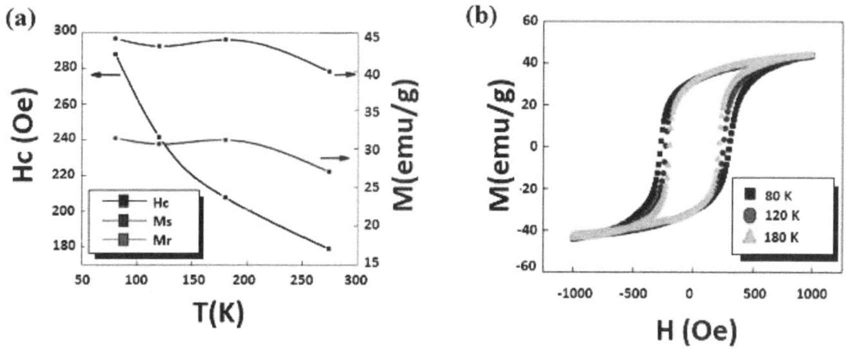

Figure 11. (a) Low-temperature VSM measurements of Hc, Ms and Mr measured at several temperatures for thin films of Fe/MgO/Fe synthesized at 100°C. The lines are drawn connecting the points as a guide. (b) Hysteresis loops of Fe/MgO/Fe/MgO (100) thin film synthesized at 100°C measured at several low-temperature values.

5. Discussion

As shown in Figure 5, the in-plane coercive field (Hc) of nanowires is higher than thin films' Hc by 754%, and nanocolumns' Hc is higher than thin films' Hc by 403%. These higher values of Hc for nanowires are due to magneto crystalline shape anisotropy since both nanowires' and nanocolumns' Hc values are higher than Hc of planar films Magneto crystalline anisotropy according to density functional theory is due to change in the relative occupancy of the 3d orbitals of Fe atoms at the interface of Fe/MgO [15]. The reason nanowires' Hc is higher than nanocolumns' Hc is encapsulation by carbon nanotubes and subsequent interaction of C atoms with Fe atoms. This interaction between C and Fe atoms is between the π-electronic states of carbon and 3d bands of the Fe surface. First-principle calculation predicts weak interaction between the Fe layer and the MgO substrate making the Fe film to act as a free-standing Fe monolayer (3.10 μB), with enhanced magnetic moment [13]. Mössbauer measurement has also shown higher hyperfine field attributed to the interface region between the epitaxial Fe and MgO layers [16]. This study on the hybrid interface between carbon-based organic molecules and ferromagnetic surfaces is very important in the development of wearable spintronics and environmentally friendly sensors based on organic spintronics [17]. Nanowires of α-Fe synthesized inside alumina templates and single nanowires inside dense nickel nanowire arrays [18, 19] depict a B-H loop that narrows when the field is perpendicular to the nanowire axis; this deformation in the B-H loop indicates small dipole interactions. The small dipole interactions in these systems are due to the large spacing between the nanowires. In this chapter, nanowires of Fe/MgO/Fe for all orientations of the applied field depict B-H loops that have the same width without narrowing deformation in the B-H loops indicating high dipole interactions. The reason for this high dipolar interactions is the densely grown nanowires with close proximity to each other.

6. Conclusions

In conclusion, this chapter reports on the magnetic study of Fe/MgO/Fe of planar nanometric thin films, nanocolumns and nanowires encapsulated inside carbon nanotubes. Deposition conditions in particular geometry and most notably encapsulation influence both structural and magnetic properties of Fe/MgO/Fe TMR. Nanowires of Fe/MgO/Fe showed superior magnetic properties than planar thin films. The in-plane value of the coercive field of nanowires is higher by 754% than planar thin films.

References

1. Wolf S, Awschalom D, Buhrman R, Daughton J, Von Molnar S, Roukes M, Chtchelkanova AY, Treger D. Spintronics: A spin-based electronics vision for the future. Science. 2001;294:1488-1495
2. Butler WH, Zhang XG, Schulthess TC, MacLaren JM. Spin-dependent tunneling conductance of Fe/MgO/Fe sandwiches. Physical Review B. 2001;63:054416
3. Waldron D, Timoshevskii V, Hu Y, Xia K, Guo H. First principles modeling of tunnel magnetoresistance of Fe/MgO/Fe trilayers. Physical Review Letters. 2006;97:226802
4. Mathon J, Umerski A. Theory of tunneling magnetoresistance in a disordered Fe/MgO/Fe (001) junction. Physical Review B. 2006;74:140404
5. Fan Y, Smith KJ, Lupke G, Hanbicki AT, Goswami R, Li CH, Zhao HB, Jonker BT. Exchange bias of the interface spin system at the Fe/MgO interface. Nature Nanotechnology. 2013;8:438-444
6. Yuasa S, Nagahama T, Fukushima A, Suzuki Y, Ando K. Giant room-temperature magnetoresistance in single-crystal Fe/MgO/Fe magnetic tunnel junctions. Nature Materials. 2004;3:868-871
7. Parkin SS, Kaiser C, Panchula A, Rice PM, Hughes B, Samant M, Yang S-H. Giant tunnelling magnetoresistance at room temperature with MgO (100) tunnel barriers. Nature Materials. 2004;3:862-867
8. Newman A, Khatiwada S, Neupane S, Seifu D. Nanowires of Fe/multi-walled carbon nanotubes and nanometric thin films of Fe/MgO. Journal of Applied Physics. 2015;117:144302
9. lópez-Urías F, Munoz-Sandoval E, Reyes-Reyes M, Romero A, Terrones M, Morán-López J. Creation of helical vortices during magnetization of aligned carbon nanotubes filled with Fe: Theory and experiment. Physical Review Letters. 2005;94:216102

10. Mauge-Lewis K, Mojibola A, Toth EA, Mohammed M, Seifu D, Aslan K. Metal-assisted and microwave-accelerated evaporative crystallization: Proof-of-principle application to proteins. Crystal Growth & Design. 2015;7:3212-3219

11. Wei B, Vajtai R, Jung Y, Ward J, Zhang R, Ramanath G, Ajayan P. Assembly of highly organized carbon nanotube architectures by chemical vapor deposition. Chemistry of Materials. 2003;15:1598-1606

12. Aryee D, Seifu D. Shape anisotropy and hybridization enhanced magnetization in nanowires of Fe/MgO/Fe encapsulated in carbon nanotubes. Journal of Magnetism and Magnetic Materials. 2017;429:161-165

13. Li C, Freeman A. Giant monolayer magnetization of Fe on MgO: A nearly ideal two-dimensional magnetic system. Physical Review B. 1991;43:780

14. Luches P, Benedetti S, Liberati M, Boscherini F, Pronin II, Valeri S. Absence of oxide formation at the Fe/MgO (001) interface. Surface Science. 2005;583:191-198

15. Niranjan MK, Duan C-G, Jaswal SS, Tsymbal EY. Electric field effect on magnetization at the Fe/MgO (001) interface. Applied Physics Letters. 2010;96:222504

16. Balogh J, Dézsi I, Fetzer C, Korecki J, Kozioł-Rachwał A, Młyńczak E, Nakanishi A. Magnetic properties of the Fe-MgO interface studied by Mössbauer spectroscopy. Physical Review B. 2013;87:174415

17. Wong PKJ, Tran TLA, Brinks P, van der Wiel WG, Huijben M, de Jong MP. Highly ordered C60 films on epitaxial Fe/MgO (001) surfaces for organic spintronics. Organic Electronics. 2013;14:451-456

18. Peng Y, Zhang H-L, Pan S-L, Li H-L. Magnetic properties and magnetization reversal of α-Fe nanowires deposited in alumina film. Journal of Applied Physics. 2000;87:7405-7408

19. Nielsch K, Hertel R, Wehrspohn R, Barthel J, Kirschner J, Gösele U, Fischer S, Kronmüller H. Switching behavior of single nanowires inside dense nickel nanowire arrays. IEEE Transactions on Magnetics. 2002;38:2571-2573

2

Diamond Nanowire Synthesis, Properties and Applications

1. Introduction

The development of nanowires for the effective applications still seems to be interesting and a challenging task [1, 2, 3]. So far, many kinds of nanowires have been constructed from various sources such as metals, polymers, inorganic-organic hybrid systems and semiconductor materials [4, 5, 6, 7]. Those nanowires have the diverse mechanical, electrical, thermal and multifunctional properties [8, 9]. Further, nanowires can be effectively utilized in electrical transport studies, electrochemical studies and solar energy conversions [10, 11]. In this footpath, diamond nanowires (DNWs) are also known as a material of extremes, in which, its properties are exceptional in terms of band gap, electron affinity, chemical inertness, resistance to particle bombardment, hardness and thermal conductivity [12]. Moreover, upon tuning the n- or p-doping on DNWs, the diverse field emission, semiconductor and sensory applications can be attained. The diverse applications may be attributed to the lattice structures of those DNWs as well as the carbon-carbon bond or existence of sp2/sp3 ratio [13, 14]. However, the presence of defects such as nitrogen vacancy center (NV−) and impurity channels also enhances the benefits of DNW-based extensive research [15]. As a consequence of those defects or impurity channels, DNWs have the color centers, which enable their photonic applications such as single-photon emission [16]. Moreover, the toughness and wear resistance of DNWs may be enhanced due to the hindering of dislocation movement by the impurities.

Attributed to the utilities of DNWs, numerous reports on their synthesis have been available so far. However, the synthesis of DNWs was claimed to be a low probability event in terms of reproducibility, which makes it as a thought-provoking task. Therefore, researchers tend to develop the suitable methods to grow the DNWs due to its potential benefits in ultraviolet (UV) light detectors and emitters [17, 18] radiation particle detectors [19], high-speed and high-power field effect transistors

[20], field emission sources [21, 22], position-sensitive biochemical substrates [23] and room temperature-stabilized high-efficiency single-photon emitters [24]. So far, DNWs were grown from (1) Plasma-assisted reactive-ion etching process (RIE) with mask and maskless techniques; (2) chemical vapor deposition (CVD) techniques with diverse templates assistance, plasma enhancement, catalyst assistance, and so on; (3) sp2-hybridized carbon and sp3-hybridized diamondoids (post-treatment of multiwalled carbon nanotubes (MWCNTs) with hydrogen plasma and from fullerenes). Similar to the above reported techniques on DNW growth, the development of hybrid graphene-DNWs (G-DNWs) also attracted the modern scientific research because of their diverse conductivity or semiconductor applications [12]. However, such nanowire (DNWs and G-DNWs) growth is still a challenging task; hence, an overview on its synthesis, structures and applications is required. In this chapter, we tend to present a brief report of the diamond nanowires, with discussions on DNW synthesis along with their structures, properties and applications. Wherein, the important synthetic pathways to grow the DNWs are pinpointed. Then, the comprehensive discussions on the structures and properties of the DNWs are derived from the available theoretical and experimental reports. Subsequently, the applications of those DNWs in diverse fields are summarized.

2. Synthetic strategies for DNW growth

2.1. Plasma-assisted reactive-ion etching

Reactive-ion etching (RIE) is an etching technology applied in micro- or nanofabrication, which may apply the dry etching than that of wet routes [25]. Wherein, plasma has been used to remove material deposited on substrates. A schematic diagram presented in Figure 1 represents a typical RIE setup.

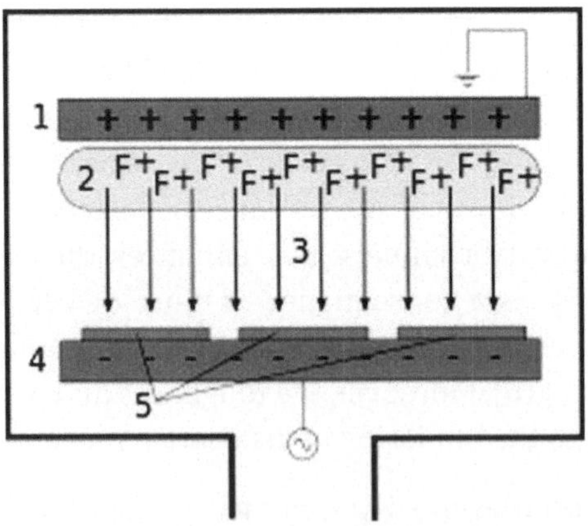

Figure 1. A diagram of a common RIE setup. An RIE consists of two electrodes (1 and 4) that create an electric field (3) meant to accelerate ions (2) toward the surface of the samples (5) (https://en.wikipedia.org/wiki/Reactive-ion_etching).

In general, the plasma is generated by an electromagnetic field under vacuum. Then, plasma produces the high-energy ions, which react with the surface of the sample to provide the desired nanostructures. However, its output also depends on the parameters such as power density, frequency, pressure, dc bias, gas composition, flow rate and so on. In 1997, Shiomi et al. informed the DNW growth by means of plasma-assisted RIE technique [26]. Thereafter, RIE technique has been widely applied for the growth of DNWs. However, later on, DNWs were grown up either with the support of mask or maskless processes.

2.1.1. RIE with masks for DNW growth

The planar diamond films can be etched to obtain the DNWs. This also was attained with the support of several masks such as (1) metal nanoparticles mask, (2) oxide nanoparticles mask, and (3) diamond nanoparticles mask. However, the size and density of the developed DNWs depends on the nanoparticles that are used in masks, in which the size of those nanoparticles may lie in few nanometers.

(A) RIE with metal mask: After Shiomi's report [26] on DNW growth by using Al mask and oxygen plasma RIE, various columnar diamond nanowires with 300-nm length and 10-nm width have been constructed through etching CVD polycrystalline diamond films in O2 plasma [27, 28]. In this light, Liao et al. effectively developed the single-crystal diamond pillar like DNWs by Al-masked RIE technique [29]. However, Al-masked RIE method led to provide polycrystalline DNWs, hence having the disadvantages such as the presence of grain boundaries, impurities and large stresses in the films. Apart from Al-mask, other kinds of metals such as Mo, Ni, Fe and Au were also been utilized to develop doped or undoped DNWs [30, 31, 32]. Li and Hatta explored the effect of those metal masks, for the development of DNWs [33].

(B) RIE with oxide nanoparticles mask: Fujishima et al. successfully developed the DNWs through reactive-ion etching supported by oxygen plasma consisting of two-dimensional (2D) arrays of monodisperse solid SiO2 particles as masks [34]. Wherein, on the planar diamond surface, fine SiO2 particles are packed at high density [35] and oriented layers over a wide surface area by water evaporation and lateral capillary forces [36]. Then, reactive-ion etching (RIE) was carried out with oxygen plasma through the SiO2 arrays for 5–120 min in a plasma-etching machine with a radio frequency (RF) generator. Lastly, the SiO2 particles were detached from the diamond by HF-HNO3 treatment, which afford the DNW arrays. After this report, Hausmann et al. also elaborate the DNW synthesis by Al2O3 mask [37], which found to be the most etch resistant. Hence, these flowable oxide masks are demonstrated to be a suitable etching mask for the construction of ordered arrays of DNWs.

(C) RIE with diamond nanoparticles: As shown in Figure 2, by using diamond nanoparticles as a mask, Yang et al. described the vertically aligned DNW synthesis from boron-doped single-crystalline CVD diamond films [38]. Initially, a microwave-assisted CVD technique is used to grow the boron-doped (P-type) diamonds with smooth surfaces by homoepitaxy on Ib diamond substrates. Next, diamond nanoparticles etching mask with well-defined size and quality is deposited. The size of those diamond nanoparticles (dissolved in water by ultrasonication to form a pseudostable suspension) lies between 8 and 10 nm. Thereafter, to seed diamond nanoparticles on the surface of a diamond substrate, the planar diamond film is immersed into the suspension and sonicated. After deposition, RIE in an O2 (97%)/ CF4 (3%) gas mixture is applied to afford the vertically aligned DNWs, which has been utilized in DNA sensing [39].

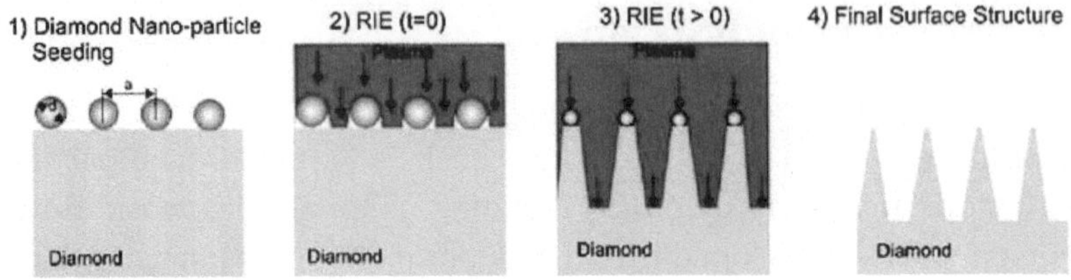

Figure 2. Schematic illumination of the fabrication of vertically aligned diamond nanowires using a top-down technology and using diamond nanoparticles as the etching mask. Reproduced with permission from [38].

2.1.2. Maskless RIE for DNW growth

The above mask methods have certain limitations and hence become unfavorable for large-scale fabrication. Therefore, researchers tend to develop uncomplicated methods to remove some masks by additional chemical or physical processes that can grow the DNWs. In this way, Fujishima et al. described the synthesis of heavily B-doped DNWs (the boron doping level is 2.1 × 1021 B cm−3) through oxygen plasma without any additional mask [40]. Here, boron atoms on the diamond act as the mask during plasma etching, hence avoiding the deposition or removal of mask by additional steps. In detail, during the etching step, those boron oxide species are removed collectively with carbon atoms, and then they appear to redeposit near the tops of the DNWs, which may serve as an etching mask. This straightforward maskless method has been widely used for the synthesis of DNWs in recent times.

2.2. Chemical vapor deposition (CVD) for DNW synthesis

Among the available effective methods for DNW synthesis, CVD technique is one of the promising processes utilized extensively [41]. This simple evaporation technique has been used to grow the elemental or oxide nanowires in an appropriate atmosphere. However, CVD process can be applied by means of template assistance or template-free ones as follows.

2.2.1. Template-assisted CVD methods for DNW growth

This is a convenient method to generate the 1D nanostructures and capable of producing nanostructures with exclusive structures, morphologies and properties [42, 43, 44]. Wherein, the template assists as a scaffold on which other materials with similar morphologies are produced. Moreover, they can be at nanoscale within mesoporous alumina or polycarbonate membranes. The following templates were applied so far to grow DNWs: (A) nanowires templated with CVD and (B) anodic aluminum oxide (AAO) templated CVD.

(A) Nanowires templated CVD for DNW growth: This method has two steps such as (1) synthesis of various nanowire templates and (2) conformal coating of nanowires templates with nanodiamond, which may lead to form the DNWs by CVD technique. Firstly, May et al. explored the microdiamond coatings into tungsten wires through CVD to construct the DNWs [45]. Afterward, several researchers applied this technique by using silicon, carbide, tungsten, titanium and copper nanowires as a template for DNW synthesis. Figure 3 illustrates such B-doped DNW growth by Si nanowires template [46].

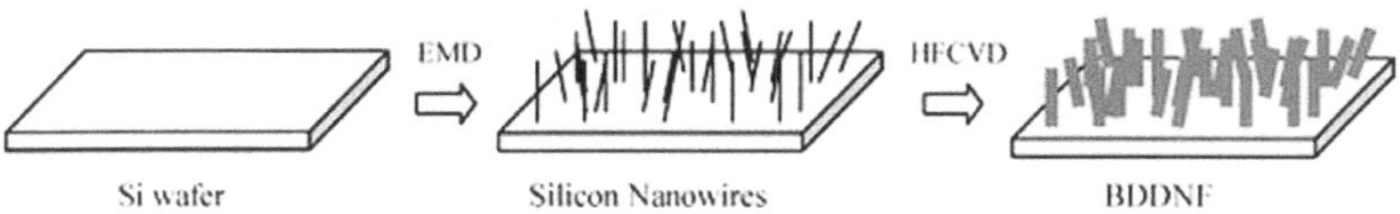

Figure 3. Fabrication of B-doped DNWs by Si-nanowires templates with CVD. Reproduced with permission from [46].

(B) Anodic aluminum oxide (AAO) templated CVD: Masuda and coworkers illustrated the growth of polycrystalline DNWs and diamond-like carbon (DLC) nanotubes by means of anodic aluminum oxide templates in microwave plasma-assisted CVD [47], in which those alumina templates [48] were prepared by electrochemical anodization of an aluminum sheet. Alumina templates possessing the holey nanoporous membranes and nucleated with 50- and 5-nm diamond particles led to the formation of DNWs. In this process, the deposition of diamond through the alumina pores yields a continuous film and supports the formation of nanostructures. Finally, by immersing in concentrated phosphoric acid at 250°C, those nanostructures can be released from the alumina.

2.2.2. Template-free CVD techniques for DNW growth

More recently, template-free CVD methods for DNW synthesis attracted the scientific community. Those template-free CVD techniques are (A) microwave plasma-enhanced CVD (MPCVD), (B) hot cathode direct current plasma CVD (HCDC-PCVD), and (C) catalyst-assisted atmospheric pressure CVD. Detailed information of the above CVD methods is presented subsequently.

(A) Microwave plasma-enhanced CVD (MPCVD): At first, Valsov et al. presented the synthesis of hybrid graphite-diamond nanowires (G-DNWs) over an ultrananocrystalline diamond (UNCD) film by using MPCVD technique [49]. Afterward, Shang and coworkers described the development of ultrathin diamond nanorods (UDNRs) by this method [50]. However, the incorporation of N2 becomes essential as demonstrated by recent reports [51, 52], wherein the incorporation of N2 enhances the electrical conductivity through tuning the sp2/sp3 carbon ratio. Hypothetically, the introduction of nitrogen into plasma may motivate the formation of molecular CN species, thereby generating favorable conditions for an increase in the grain size as well as the formation of ID diamond nanostructures. In general, DNWs synthesized through MPCVD technique show good electrochemical properties due to the rise of sp2 content, new C-N bonds at the grains and an escalation in the electrical conductivity at the grain boundaries [53].

(B) Hot cathode direct current plasma CVD (HCDC-PCVD): This is an innovative technique for the deposition of nano- and micro-crystalline diamond films with uniformity over a large area and with a high growth rate. Here, the cathode made up of a tantalum disc linked to a water-cooled cylindrical copper block, water-cooled copper block anode and a nonpulsed-type dc power source is used. From this technique, Zeng et al. explored the formation of DNRs along with (111) diamond microcrystals and (100) diamond microcrystals on Si substrates [54].

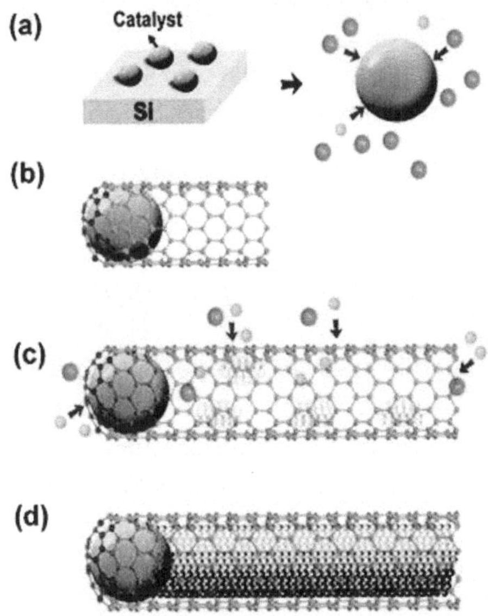

Figure 4. Schematic diagram showing a possible formation process of diamond nanowires: (a) catalytic particles are formed from the evaporated or deposited thin film on the substrate as the temperature rises; (b) carbon-containing radicals reach the surface of the catalysts, leading to the growth of either a diamond stud or a graphitic tube via the VLS mechanism. The size is determined by the catalyst. (c) Hydrogen assists in the growth process by either preferentially etching sp. and sp2 bonds or transforming them into sp3 bonds. With the higher capillary pressure at smaller diameters, the diamond phase could be more stable but the capillary pressure rapidly decreases with diameter, leaving the shell more stable in the graphitic phase. (d) The grain growth, boundary healing and structural reorganization take place in the slow-cooling period in the presence of a pure hydrogen flow. Reproduced with permission from [57].

(C) Catalyst-assisted atmospheric pressure CVD: Apart from the previously mentioned high-temperature methods assisted by plasma or energy radiation, the production of long single-crystalline DNWs by conventional thermal CVD methods has become essential because of their potential benefits. In this light, Hsu et al. described the growth of DNWs by means of CVD without plasma or energy sources and at atmospheric pressure [55]. Here, methane and hydrogen were flowed into the Fe catalyst solution, which was dispersed on an Si substrate at 900°C. Then, pure hydrogen was run through the quartz tube chamber (at 200 sccm as rate) without pumping the residual methane. Subsequently, the temperature lowers down to an ambient condition at a rate of 1.2°C min−1 for 12 h. This method will produce the uniform long and thin DNWs with a diameter of 60–90 nm. Importantly, in this process, hydrogen plays a vital role in the formation of DNWs via sp- and sp2-hybridized bonds transformation into sp3-hybridized atoms [56]. Figure 4 demonstrates the possible vapor-liquid-solid (VLS) mechanism for the growth of DNWs by this method [57]. This technique supports the exceptional utilization CVD with the support of transition metal catalyst such as Fe and also envisioned the applicability of CVD at atmospheric pressure.

2.3. DNW growth from sp2 carbon and sp3 diamondoids

Attributed to the importance, the conversion of sp2 graphite carbon to sp3 diamond crystals remains to be a challenging task over many years for which high pressures and high temperatures are required. Recently, researchers developed few methods for such transformation; however, it is still a challenging task. The following are few examples of transformation of sp2 graphite carbon to sp3 carbon, which led to the formation of DNWs: (A) DNW synthesis by hydrogen plasma post-treatment of multiwalled carbon nanotubes, (B) DNW growth from fullerenes and (C) DNWs from diamondoids.

(A) DNW synthesis by hydrogen plasma post-treatment of multiwalled carbon nanotubes (MWCNTs): In 2005, Sun and coworkers presented this simple method for the growth of DNWs from carbon nanotubes (CNTs) via hydrogen plasma post treatment [58]. Impressively, upon extended hydrogen plasma treatment, the DNWs with the diameters of 4–8 nm and with the lengths up to several hundreds were obtained. This work also revealed the TEM of single-crystal DNWs from the MWCNTs after treatment in hydrogen plasma at 1000 K for 20 h. The author proposed the mechanism as clustering, crystallization, growth and faceting, which is similar to the report by Singh et al. [59]. It is also established that the presence of amorphous carbon sheath over diamond nanoparticles and DNWs is responsible for this kind of transformation.

(B) DNW growth from fullerenes: In 2005, Dubrovinskaia and coworkers synthesized a bulk sample of nanocrystalline cubic diamond from fullerene C60, which have the crystallite sizes of 5–12 nm and a hardlike single-crystal diamond

[60]. Figure 5 represents the TEM of those ADNRs. These nanocrystalline diamonds seem to be highly stable at an elevated temperature and an ambient pressure. In the meantime, they developed the aggregated diamond nanorods (ADNRs) from C60 by multi-anvil apparatus [61]. Those ADNRs have the diameter of 5–20 nm and have the length of more than 1 μm.

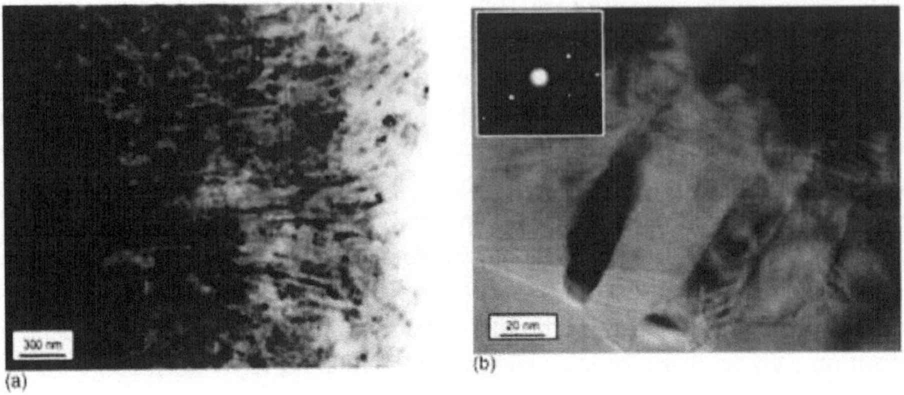

Figure 5. (a) Bright-field TEM image of a nanocrystalline aggregate with needle-shaped, elongated crystals diamond nanorods. The crystals can be longer than 1 μm, whereby the needle width is only about 20 nm or less; (b) bright-field image shows a close-up of the elongated crystals. The long edges of the crystals are parallel to the (111) plane, and the needle axes are approximately parallel (211)*. Reproduced with permission from [60].

(C) DNWs from diamondoids: Similar to carbon nanotubes and fullerenes, diamondoids may also lead to the formation of DNWs. The 1D diamondoid aggregates confined in CNTs directed to form the DNWs via 'face-fused' reaction. However, these transformations of adamantane into DNWs seem to be energetically not feasible. Contrarily, Zhang et al. explored the theoretical and experimental proof for these fusion reactions by diamantane-4,9-dicarboxylic acid transformation to 1D diamond nanowires inside CNTs [62]. In which, the fusion of diamantane-4,9-dicarboxylic acid under the confinement of CNTs yields the DNWs.

2.4. Wet chemical route to synthesis DNWs

Attributed to the applications of DNWs, numerous efforts have been made by the researchers to synthesize them. Among them, wet chemical route seems to be impressive with respect to cost-effectiveness than that of RIE and CVD techniques. But it is also essential to make them with reproducibility and uniformity. To this footpath, recently, our group report the pH-induced electrostatic self-assembly of novel cysteamine functionalized diamond nanoparticles (ND-Cys) to evidence hybrid G-DNW growth [63]. Those G-DNWs are highly stable in respective pH buffers, but if more amount of DI-water is added, the longer nanowires (initially at ~100 μm) break into small wires/rods (few microns). At pH 6, the width of G-DNWs ranges between

20 and 800 nm and the length lies between 200 nm and hundreds of microns with respect to dispersion concentration. Wherein, the DNW formation was initiated through electrostatic forces within the partially graphitized ND-Cys particles. Next, those partially graphitized ND-Cys particles and defects/impurity channels were further promoted to form the graphene shells on the surface of DNPs and sandwiched between the diamond cores. These G-DNWs show exceptional conductivity due to the presence of defects and impurity channels. Figure 6 illustrates the TEM image of those nanowires with defect or impurity channels.

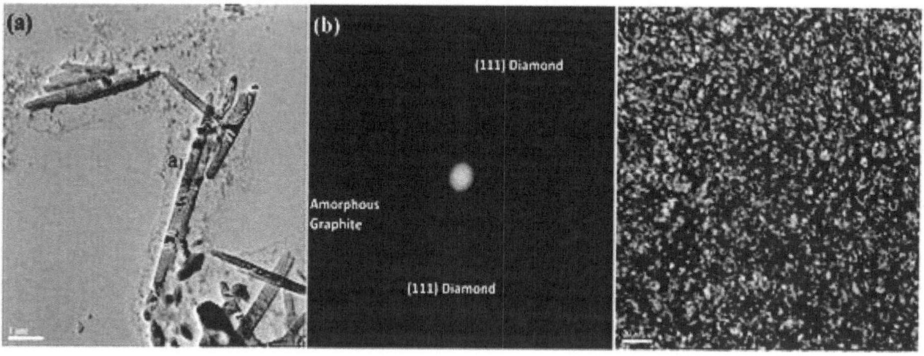

Figure 6. (a) HR-TEM image of G-DNWs, (b) FT pattern of selected area a1 representing amorphous graphite along with diamond (111) diffraction pattern and (c) high magnification image of a1 region representing less perfect graphite layer along with defects or impurity channels. Reproduced with permission from [63].

In this way, with respect to Berman et al. report on metal-induced graphitization of diamond particles [64], metal ions induced G-DNWs formation is also seem to be highly feasible. However, the reproducibility and percentage formation of G-DNWs by this path is still a challenging task. Currently, our group is working on this research to grow the G-DNWs with good reproducibility.

3. Structures and properties of DNWs

In order to establish the diverse applications of DNWs, the structure and properties should be elucidated. The properties such as structural stability, mechanical properties, density and compressibility, photon optical mode and electronic structure, thermal conductivity and electrochemical properties play vital role in their applications. Hence, researchers described the experimental and theoretical investigations on the structure and properties of DNWs as follows.

3.1. Structural stability of DNWs

From theoretical investigations, it has been found that dehydrogenated C(111) octahedral nanodiamond surfaces are structurally unstable. However, cuboctahedral structures of nanodiamond may increase the C(100) surface area and become more stable, which also reduce the surface graphitization. In this light, Barnard et al. investigated three kinds of DNWs including dodecahedral, cubic and cylindrical nanowires and found that nanocrystalline diamonds are structurally stable at one dimension [65]. Moreover, they also demonstrate that stability depends on the surface morphology and crystallographic direction of the principal axis of DNWs. In a similar fashion, Tanskanen and coworkers established the structures of polyicosahedral DNWs derived from diamondoids, C20H20, C20@C80H60, and C20@C80@C180H120. For which they have summarized the HOMO-LUMO gaps, and band gaps via B3LYP calculations [66]. Wherein, the C20@C80@C180H120 structures are energetically favored and the DNWs at 110 direction have the lowest strain energies leading to more stability. This has been experimentally proved by the stability of DNRs (at 110 direction) synthesized through hydrogen plasma post-treatment of multiwalled CNTs [58], whereas the DNWs at 100 direction seem to be unstable as reported earlier [67].

3.2. Mechanical properties of DNWs

Tanskanen et al. described the mechanical properties of DNWs through Poisson's ratios, Young's moduli and shear moduli interrogations [66], which proved that (111) DNWs have the highest Young's moduli than the (110) and (111) DNWs. In this report, they suggest that polyicosahedral DNWs have more strain than that of conventional DNWs. In this way, Guo and coworkers presented the mechanical properties of (001) DNWs by means of molecular dynamics simulations [68] and specified that Young's modulus of those DNWs is lower than those of bulk diamond. Similarly, Jiang et al. explored Young's modulus of DNWs in different crystallographic orientations as a function of cross-sectional area [69]. Wherein, Young's modulus has the sequence of (100), (110), (111) and (112) directions and indicated that those values are lower than the bulk value and increase with its cross-sectional area.

3.3. Density and compressibility of DNWs

Initially, Dubrovinskaia and Dubrovinsky reported the density of the aggregated diamond nanorods (ADNRs), which were developed from fullerene C60 by multi-anvil apparatus [61]. The X-ray density of ADNRs is about 0.2–0.4% greater than the bulk diamond, which also corresponds to the measured density of 3.532(5) g cm−3. The higher density of ADNRs may arise from the outerlayer contraction leading to shortening of the C-C bonds inside the diamond. In this work, they have also evaluated the compressibility of ADNRs by using the third-order Birch-Murnaghan equation of state, wherein they established the >11% lesser compressibility of ADNRs than that of usual diamond.

3.4. Phonon optical modes and electronic properties of DNWs

Trejo and coworkers reported the optical phonons and Raman-scattering properties of DNWs by using a local bond polarization model based on the displacement–displacement Green's function and the Born potential [70]. Further, they have also studied the electronic band structure of DNWs through a semiempirical tight-binding approach and compared with density functional theory (DFT) studies. From the calculations, they have concluded that phonons and electrons tend to show a clear quantum confinement signature. Moreover, this study also establishes that during the DNWs width increase, the Raman peak shifts to lower frequencies due to the phonon confinement, as reported by our group [71]. Subsequently, the band gap also decreases as the width of the DNWs increases.

3.5. Thermal conductivity and electrochemical properties of DNWs

In general, it is recognized that the thermal conductivity of DNWs may not be incredibly affected by surface functionalization. However, at nanometer scale, dimensions of DNWs may reduce the thermal conductivity than that of bulk diamond as demonstrated by Novikov et al. [72]. In this way, Moreland and coworkers explored that the conductivity of DNW is lower than the CNT and depends on the choice of thermostat [73]. Similarly, Guo et al. described that the thermal conductivity of DNWs may rise with respect to the increase in length and cross-sectional areas [74]. Recently, as seen in Figure 7, our group also proved the downfall in the conductivity of a single G-DNW with respect to a decrease in temperature [63]. Overall, it has been concluded that between 0 and 1000 K, DNW's thermal conductivities firstly upsurge with an increasing temperature and then dropdown.

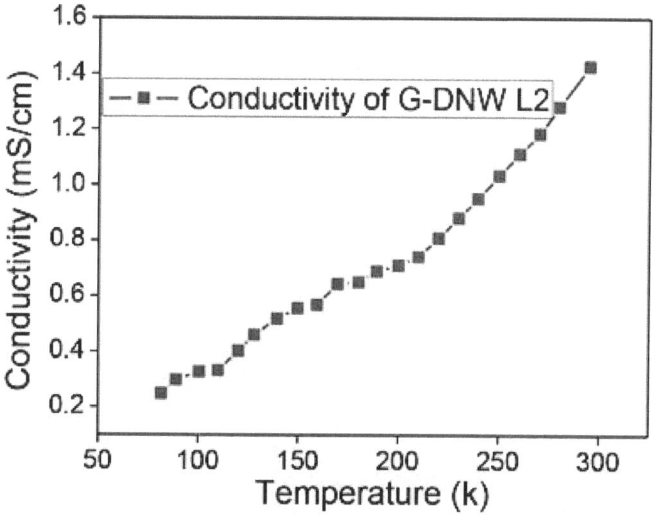

Figure 7. Temperature-dependent conductivity of DNW L2. Reproduced with permission from [63].

Next, coming to the electrochemical properties, it is well recognized that the planar boron-doped diamond (BDD) materials have the unique physical properties and were already been effectively applied as electrodes in many sensing studies. Wherein, compared to glassy carbon electrode, the diamond electrode acts as a potential candidate due to its chemical stability and biocompatibility [75]. Moreover, BDD electrode is not fouled easily and has a low background current with a wide potential window. By altering the surface end of BDD, the electronic and chemical properties can be tuned according to the requirement. Currently, the BDD nanograss array is also involved in electron transport and electrocatalytic utilities [76, 77]. Conclusively, it is well established that the nanotextured DNW surfaces become the suitable platform for novel biosensor investigations.

4. Applications of DNWs

Among the applications of DNWs, the following five utilities have been demonstrated strongly. Those applications are (1) field emission applications of DNWs, (2) DNWs in mass analysis of small molecules, (3) DNWs as nanoelectromechanical switches, (4) DNWs as electrochemical sensors and (5) DNWs in ultrasensitive force microscopy.

4.1. Field emission applications of DNWs

The negative electron affinity of DNWs has been used in field emission studies. Recently, reports on the electron field emission (EFE) properties of CVD-developed ultracrystalline diamond and hybrid diamond-graphite films were reported [78, 79]. In this way, the EFEs of DNWs were also been described by (A) planar DNWs array and (B) single DNW.

(A) Electron field emission of planar DNW array: Lee et al. demonstrated the EFE characteristics of planar diamond film array, which has been developed by CVD techniques [80]. Recently, Sankaran et al. presented the improved EFE applications of graphite-wrapped DNWs [81]. In this path, the above group determined the enhanced electron field emission of vertically aligned ultrananocrystalline diamond needles via ZnO coating to form the heterostructured nanorods [82]. Wherein, it shows a high emission current density of 5.5 mA cm−2 at 4.25 V μm−1 and has a low turn-on field of 2.08 V μm−1 than that of bare Zn-nanorods. This outstanding emission property of planar diamond film arrays seems to be impressive to apply as the electron emitters in flat display panels.

(B) Electron field emission of a single DNW: Recently, Hsu and coworkers presented the electron field emission of a single DNW [57]. Wherein, the threshold field of DNW (1.25 Vμm−1) is four times lower than that of carbon nanotube (5 Vμm−1). This might be due to the electron affinity of DNW and defects existence. In addition, the EFE property of DNW may be attributed to its chemical inertness, high mechanical strength and high thermal conductivity.

4.2. DNWs in mass analysis of small molecules

Firstly, Coffinier et al. described the matrix-free laser desorption/ionization (D/I) mass spectrometric utilization of boron-doped DNWs (BDD NWs) toward small molecular analysis [83], in which the S/N ratios of UDD NWs are very low than that of BDD NWs. Therefore, the potentiality of BDD NWs in mass analysis of small molecules has been proved.

4.3. DNWs as nanoelectromechanical switches

Recently, researchers tend to develop the diamond-based nanoelectromechanical (NEM) switches as an alternative silicon-based ones due to its exceptional properties such as high Young's modulus, maximum hardness, hydrophobicity, low mass density, greater thermal conductivity, extraordinary corrosion resistance and low toxicity. However, because of existed grain boundaries, impurities, large stress, low electrical conductivity and poor reproducibility, the polycrystalline or nanocrystalline film-based switches seem not to be as impressive candidates [84, 85]. In contrast, the utilization of single-crystalline DNWs as NEM is appraised by Liao and coworkers [29], in which those switches show low leakage current (<0.1 pA) with a high ON/OFF ratio, hence can compete Si-NEMS structures.

4.4. DNWs as electrochemical sensors

DNWs were effectively applied in many electrochemical sensory studies. For example, Yang and Nebel utilized the vertically aligned diamond nanowires toward DNA detection via electrochemical approach [86]. Figure 8 represents the schematic of biofunctionalized vertically aligned diamond nanowires for the determination of DNA in the abovementioned report.

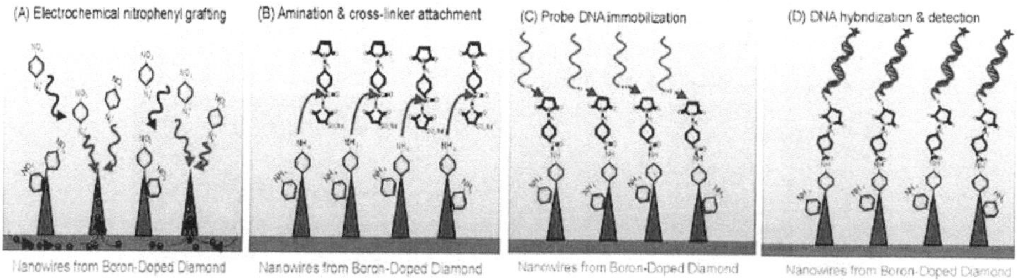

Figure 8. Schematic illumination of the biofunctionalization of vertically aligned diamond nanowires to realize a nanoscaled spacing between DNA molecules. Reproduced with permission from [86].

Further, they have protracted those diamond nanowires in electrochemical gene sensors [87]. Akin to vertically aligned nanowires, BDD NWs were applied in nonenzymatic amperometric glucose biosensing by Zhi et al. [46]. Wherein, the selective determination of glucose has been demonstrated in the presence of ascorbic acid (AA) and uric acid (UA). Meanwhile, BDD NWs were also been efficiently used in the electrochemical identification of tryptophan by Szunerits and coworkers [88]. Alternatively, Lee and Lin collaborators developed a nitrogen incorporated DNW electrode for the amperometric detection of urea and in situ detection of dopamine

[89]. Here, dopamine determination was well illustrated in the presence of AA and UA. More recently, Peng et al. reported the detection of CO gas by BDD NWs through electrochemical studies [90]. Wherein, the boron-doped ultrananocrystalline diamond (B-UNCD) nanowires (NWs) evidenced greater selectivity to CO gas than that of competitive species.

4.5. DNWs in ultrasensitive force microscopy

Recently, Tao et al. described the utility of DNWs as tips for ultrasensitive force microscopy experiments [91]. Wherein, they have fabricated two types of tips using the upper and lower halves of a DNW by means of a top-down plasma etching technique and from a single-crystalline substrate. Figure 9 demonstrates the integration of diamond nanowire tips on ultrasensitive silicon cantilevers. The typical lengths of those DNWs lie in few micrometers with diameters around 100 nm. Moreover, the tip radii were at the order of 10 nm, hence becoming suitable for scanning probe applications [32].

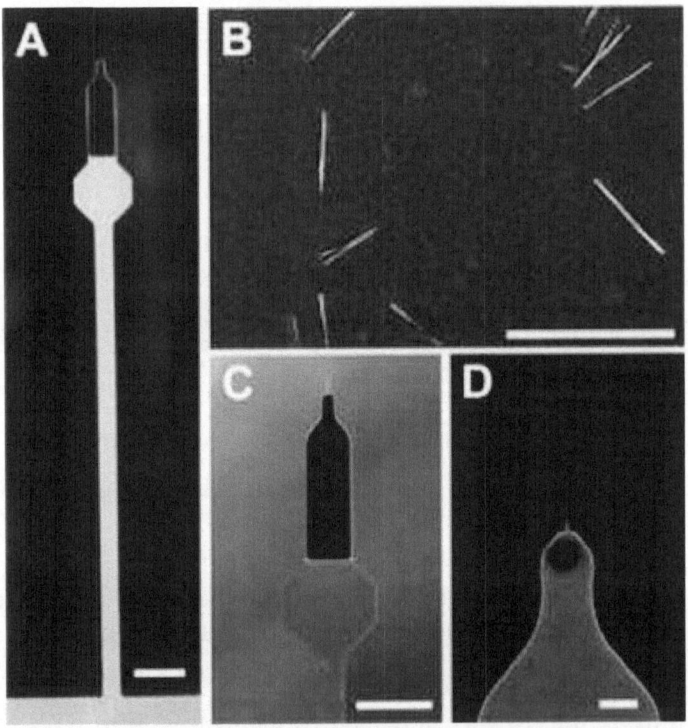

Figure 9. Integration of diamond nanowire tips on ultrasensitive silicon cantilevers. (A) Bare silicon cantilever with a nominal length of 90 μm, a shaft width of 4 μm and a thickness of 135 nm. The scale bar is 10 μm. (B) Batch of DNWs transferred onto an Si substrate for manual pickup. The scale bar is 10 μm. (C, D) zoom-in onto the end region of two different cantilevers where DNW tips had been attached. Scale bars are 10 μm in C and 1 μm in D. Reproduced with permission from [91].

5. Conclusions and perspectives

Attributed to the importance of DNWs, several efforts have been driven by experts to apply in diverse semiconductor and biological applications. In this way, those DNWs were effectively developed through different methods such as reactive-ion etching, chemical vapor deposition, from sp2 carbon and sp3 diamondoids and wet chemical route. Among them, the template-assisted synthesis of DNWs seems to be impressive to produce highly precise nanostructures. On the other hand, the cost-effective wet chemical route still remains a challenging task in terms of reproducibility and obtaining the unique structures. From experimental and theoretical studies, it has been found that DNWs have the exceptional structural, mechanical, thermal conductivity, electronic and electrochemical properties. However, structural studies on hybrid G-DNWs require exclusive focus for future applications. Subsequently, those DNWs also possess the unique applications such as EFE device, high-performance NEM switches, conductivity and electrochemical biosensor and so on. However, with respect to practicality, those applications remain unsatisfied. For instance, the reported DNW-based electrochemical biomolecules monitoring was affected by its stability; hence, still it is a challenging task to fabricate the DNW-based device for real-time continuous determination.

So far, except the wet chemical route, the reported synthetic techniques for DNWs are costly, and hence their development is still a challenging task. Therefore, much effort needed to develop the DNWs at large scale, which can be attained by the collaboration of diverse technical fields such as electro-biochemistry, nanoelectronics and analytical techniques, etc. For example, attempts are needed to develop hybrid G-DNWs by the association of CVD and wet chemical pathways. Such investigations may direct the DNWs toward diverse opto-electronic applications.

References

1. Cheng C, Fan HJ. Branched nanowires: Synthesis and energy applications. Nano Today. 2012;7:327-343. DOI: 10.1016/j.nantod. 2012.06.002
2. Liu H, Li L, Scofield ME, Wong SS. Research update: Synthesis, properties, and applications of ultrathin metallic nanowires and associated heterostructures. APL Materials. 2015;3:080701. DOI: 10.1063/1.4927797
3. Rahong S, Yasui T, Kaji N, Baba Y. Recent developments in nanowires for bio-applications from molecular to cellular levels. Lab on a Chip. 2016;16:1126-1138. DOI: 10.1039/C5LC01306B
4. Ye S, Rathmell AR, Chen Z, Stewart IE, Wiley BJ. Metal nanowire networks: The next generation of transparent conductors. Advanced Materials. 2014;26:6670-6687. DOI: 10.1002/adma.201402710
5. Lei Y, Deng P, Li J, Lin M, Zhu F, Ng T-W, et al. Solution-processed donor-acceptor polymer nanowire network semiconductors for high-performance field-effect transistors. Scientific Reports. 2016;6:24476. DOI: 10.1038/srep24476

6. Li W, Xia F, Qu J, Li P, Chen D, Chen Z, et al. Versatile inorganic-organic hybrid WO x -ethylenediamine nanowires: Synthesis, mechanism and application in heavy metal ion adsorption and catalysis. Nano Research. 2014;7:903-916. DOI: 10.1007/s12274-014-0452-9

7. Wei L, Charles ML. Semiconductor nanowires. Journal of Physics D: Applied Physics. 2006;39:R387. DOI: 10.1088/0022-3727/39/21/R01

8. Hu J, Odom TW, Lieber CM. Chemistry and physics in one dimension: Synthesis and properties of nanowires and nanotubes. Accounts of Chemical Research. 1999;32:435-445. DOI: 10.1021/ar9700365

9. Zheng G, Patolsky F, Cui Y, Wang WU, Lieber CM. Multiplexed electrical detection of cancer markers with nanowire sensor arrays. Nature Biotechnology. 2005;23:1294. DOI: 10.1038/nbt1138

10. Tian B, Zheng X, Kempa TJ, Fang Y, Yu N, Yu G, et al. Coaxial silicon nanowires as solar cells and nanoelectronic power sources. Nature. 2007;449:885. DOI: 10.1038/nature06181

11. Wanekaya A, Chen W, Myung N, Mulchandani A. Nanowire-based electrochemical Biosensors. Electroanalysis. 2006;18:533-550. DOI: 10.1002/elan.200503449

12. Yu Y, Wu L, Zhi J. Diamond nanowires: Fabrication, structure, properties, and applications. Angewandte Chemie International Edition. 2014;53:14326-14351. DOI 10.1002/anie.201310803

13. Szunerits S, Coffinier Y, Boukherroub R. Diamond nanowires: A novel platform for electrochemistry and matrix-free mass spectrometry. Sensors. 2015;15:12573. DOI: 10.3390/s150612573

14. Nagata A, Oku T, Kikuchi K, Suzuki A, Yamasaki Y, Osawa E. Fabrication, nanostructures and electronic properties of nanodiamond-based solar cells. Progress in Natural Science: Maternité International. 2010;20:38-43. DOI: 10.1016/S1002-0071(12)60004-5

15. Yonezu Y, Wakui K, Furusawa K, Takeoka M, Semba K, Aoki T. Efficient single-photon coupling from a nitrogen-vacancy center embedded in a diamond nanowire utilizing an optical nanofiber. Scientific Reports. 2017;7:12985. DOI: 10.1038/s41598-017-13309-z

16. Choy JT, Hausmann BJM, Babinec TM, Bulu I, Khan M, Maletinsky P, et al. Enhanced single-photon emission from a diamond–silver aperture. Nature Photonics. 2011;5:738. DOI: 10.1038/NPHOTON.2011.249

17. Koizumi S, Watanabe K, Hasegawa M, Kanda H. Ultraviolet emission from a diamond pn junction. Science. 2001;292:1899. DOI: 10.1126/science.1060258

18. Ito T, Nishimura M, Yokoyama M, Irie M, Wang C. Highly efficient electron emitting diode fabricated with single-crystalline diamond. Diamond and Related Materials. 2000;9:1561-1568. DOI: 10.1016/S0925-9635(00)00293-4

19. Monroy E, Omnès F, Calle F. Wide-bandgap semiconductor ultraviolet photodetectors. Semiconductor Science and Technology. 2003;18:R33. DOI: 10.1088/0268-1242/18/4/201

20. Isberg J, Hammersberg J, Johansson E, Wikström T, Twitchen DJ, Whitehead AJ, et al. High carrier mobility in single-crystal plasma-deposited diamond. Science. 2002;297:1670. DOI: 10.1126/science.1074374

21. Okano K, Koizumi S, Silva SRP, Amaratunga GAJ. Low-threshold cold cathodes made of nitrogen-doped chemical-vapour-deposited diamond. Nature. 1996;381:140. DOI: 10.1038/381140a0

22. Wang C, Garcia A, Ingram DC, Lake M, Kordesch ME. Cold field emission from CVD diamond films observed in emission electron microscopy. Electronics Letters. 1991;27:1459-1461. DOI: 10.1049/el:19910914

23. Yang W, Auciello O, Butler JE, Cai W, Carlisle JA, Gerbi JE, et al. DNA-modified nanocrystalline diamond thin-films as stable, biologically active substrates. Nature Materials. 2002;1:253. DOI: 10.1038/nmat779

24. Babinec TM, Hausmann BJM, Khan M, Zhang Y, Maze JR, Hemmer PR, et al. A diamond nanowire single-photon source. Nature Nanotechnology. 2010;5:195. DOI: 10.1038/nnano.2010.6

25. Henri J, Han G, Meint de B, Miko E, Jan F. A survey on the reactive ion etching of silicon in microtechnology. Journal of Micromechanics and Microengineering. 1996;6:14. DOI: 10.1088/0960-1317/6/1/002

26. Shiomi H. Reactive ion etching of diamond in O2 and CF4 plasma, and fabrication of porous diamond for field emitter cathodes. Japanese Journal of Applied Physics. 1997;36:7745. DOI: 10.1143/JJAP.36.7745

27. Ando Y, Nishibayashi Y, Kobashi K, Hirao T, Oura K. Smooth and high-rate reactive ion etching of diamond. Diamond and Related Materials. 2002;11:824-827. DOI: 10.1016/S0925-9635(01)00617-3

28. Mokuno Y, Chayahara A, Soda Y, Horino Y, Fujimori N. Synthesizing single-crystal diamond by repetition of high rate homoepitaxial growth by microwave plasma CVD. Diamond and Related Materials. 2005;14:1743-1746. DOI: 10.1016/j.Diamond2005.09.020

29. Liao M, Hishita S, Watanabe E, Koizumi S, Koide Y. Suspended single-crystal diamond nanowires for high-performance nanoelectromechanical switches. Advanced Materials. 2010;22:5393-5397. DOI: 10.1002/adma.201003074

30. Baik E-S, Baik Y-J, Jeon D. Aligned diamond nanowhiskers. Journal of Materials Research. 2011;15:923-926. DOI: 10.1557/JMR.2000.0131

31. Yang N, Smirnov W, Nebel CE. Three-dimensional electrochemical reactions on tip-coated diamond nanowires with nickel nanoparticles. Electrochemistry Communications. 2013;27:89-91. DOI: 10.1016/j.elecom.2012.10.044

32. Smirnov W, Kriele A, Hoffmann R, Sillero E, Hees J, Williams OA, et al. Diamond-modified AFM probes: From diamond nanowires to atomic force microscopy-integrated boron-doped diamond electrodes. Analytical Chemistry. 2011;83:4936-4941. DOI: 10.1021/ac200659e

33. Li CY, Hatta A. Preparation of diamond whiskers using Ar/O2 plasma etching. Diamond and Related Materials. 2005;14:1780-1783. DOI: 10.1016/j.diamond.2005.09.031

34. Okuyama S, Matsushita SI, Fujishima A. Preparation of periodic microstructured diamond surfaces. Chemistry Letters. 2000;29:534-535. DOI: 10.1246/cl.2000.534

35. Yamaki M, Higo J, Nagayama K. Size-dependent separation of colloidal particles in two-dimensional convective self-assembly. Langmuir. 1995;11:2975-2978. DOI: 10.1021/la00008a021

36. Dushkin CD, Kralchevsky PA, Paunov VN, Yoshimura H, Nagayama K. Torsion balance for measurement of capillary immersion forces. Langmuir. 1996;12:641-651. DOI: 10.1021/la950560p

37. 37.Hausmann BJM, Khan M, Zhang Y, Babinec TM, Martinick K, McCutcheon M, et al. Fabrication of diamond nanowires for quantum information processing applications. Diamond and Related Materials. 2010;19:621-629. DOI: 10.1016/j.Diamond2010.01.011

38. Yang N, Uetsuka H, Osawa E, Nebel CE. Vertically aligned nanowires from boron-doped diamond. Nano Letters. 2008;8:3572-3576. DOI: 10.1021/nl801136h

39. Yang N, Uetsuka H, Nebel Christoph E. Biofunctionalization of vertically aligned diamond nanowires. Advanced Functional Materials. 2009;19:887-893. DOI: 10.1002/adfm.200801392

40. Wei M, Terashima C, Lv M, Fujishima A, Gu Z-Z. Boron-doped diamond nanograss array for electrochemical sensors. Chemical Communications. 2009:3624-3626. DOI: 10.1039/B903284C

41. Cai Z, Liu B, Zou X, Cheng H-M. Chemical vapor deposition growth and applications of two-dimensional materials and their heterostructures. Chemical Reviews. 2018. DOI: 10.1021/acs.chemrev.7b00536

42. Zhang F, Wu Q, Zhang Y, Zhu J, Liu N, Yang J, et al. Chemical vapor deposition growth of InN nanostructures: Morphology regulation and field emission properties. Applied Surface Science. 2012;258:9701-9705. DOI: 10.1016/j.apsusc.2012.06.013

43. Kudo A, Steiner SA, Bayer BC, Kidambi PR, Hofmann S, Strano MS, et al. CVD growth of carbon nanostructures from zirconia: Mechanisms and a method for enhancing yield. Journal of the American Chemical Society. 2014;136:17808-17817. DOI: 10.1021/ja509872y

44. Walgraef D. Self-organization and nanostructure formation in chemical vapor deposition. Physical Review E. 2013;88:042405. DOI: 10.1103/PhysRevE.88.042405

45. May PW. CVD diamond: A new technology for the future? Endeavour. 1995;19:101-116. DOI: 10.1016/0160-9327(95)97494-S

46. Luo D, Wu L, Zhi J. Fabrication of boron-doped diamond nanorod forest electrodes and their application in nonenzymatic amperometric glucose biosensing. ACS Nano. 2009;3:2121-2128. DOI: 10.1021/nn9003154

47. Masuda H, Yanagishita T, Yasui K, Nishio K, Yagi I, Rao TN, et al. Synthesis of well-aligned diamond nanocylinders. Advanced Materials. 2001;13:247-249. DOI: 10.1002/1521-4095(200102)13:4<247::AID-ADMA247>3.0.CO;2-H

48. Masuda H, Fukuda K. Ordered metal nanohole arrays made by a two-step replication of honeycomb structures of anodic alumina. Science. 1995;268:1466. DOI: 10.1126/science.268.5216.1466

49. Vlasov I, Lebedev OI, Ralchenko VG, Goovaerts E, Bertoni G, Van Tendeloo G, et al. Hybrid diamond-graphite nanowires produced by microwave plasma chemical vapor deposition. Advanced Materials. 2007;19:4058-4062. DOI: 10.1002/adma.200700442

50. Shang N, Papakonstantinou P, Wang P, Zakharov A, Palnitkar U, Lin IN, et al. Self-assembled growth, microstructure, and field-emission high-performance of ultrathin diamond nanorods. ACS Nano. 2009;3:1032-1038. DOI: 10.1021/nn900167p

51. Shalini J, Lin Y-C, Chang T-H, Sankaran KJ, Chen H-C, Lin IN, et al. Ultra-nanocrystalline diamond nanowires with enhanced electrochemical properties. Electrochimica Acta. 2013;92:9-19. DOI: 10.1016/j.electacta.2012.12.078

52. Sobia AR, Adnan S, Mukhtiar A, Khurram AA, Turab AA, Awais A, et al. Effect of nitrogen addition on hydrogen incorporation in diamond nanorod thin films. Current Applied Physics. 2012;12:712-717. DOI: 10.1016/j.cap.2011.10.008

53. Ma KL, Zhang WJ, Zou YS, Chong YM, Leung KM, Bello I, et al. Electrical properties of nitrogen incorporated nanocrystalline diamond films. Diamond and Related Materials. 2006;15:626-630. DOI: 10.1016/j.Diamond2005.11.017

54. Zeng L, Peng H, Wang W, Chen Y, Lei D, Qi W, et al. Nanocrystalline diamond films deposited by the hot cathode direct current plasma chemical vapor deposition method with different compositions of CH4/Ar/H2 gas mixture. Journal of Physical Chemistry C. 2008;112:1401-1406. DOI: 10.1021/jp710082n

55. Hsu C-H, Cloutier SG, Palefsky S, Xu J. Synthesis of diamond nanowires using atmospheric-pressure chemical vapor deposition. Nano Letters. 2010;10:3272-3276. DOI: 10.1021/nl100616x

56. Lambrecht WRL, Lee CH, Segall B, Angus JC, Li Z, Sunkara M. Diamond nucleation by hydrogenation of the edges of graphitic precursors. Nature. 1993;364:607. DOI: 10.1038/364607a0

57. Hsu C-H, Xu J. Diamond nanowire–a challenge from extremes. Nanoscale. 2012;4:5293-5299. DOI: 10.1039/C2NR31260C

58. Sun LT, Gong JL, Zhu ZY, Zhu DZ, Wang ZX, Zhang W, et al. Synthesis and characterization of diamond nanowires from carbon nanotubes. Diamond and Related Materials. 2005;14:749-52. DOI: 10.1016/j.Diamond2005.01.025

59. Singh J. Nucleation and growth mechanism of diamond during hot-filament chemical vapour deposition. Journal of Materials Science. 1994;29:2761-2766. DOI: 10.1007/BF00356830

60. Dubrovinskaia N, Dubrovinsky L, Crichton W, Langenhorst F, Richter A. Aggregated diamond nanorods, the densest and least compressible form of carbon. Applied Physics Letters. 2005;87:083106. DOI: 10.1063/1.2034101

61. Dubrovinskaia N, Dubrovinsky L, Langenhorst F, Jacobsen S, Liebske C. Nanocrystalline diamond synthesized from C60. Diamond and Related Materials. 2005;14:16-22. DOI: 10.1016/j.Diamond2004.06.017

62. Zhang J, Zhu Z, Feng Y, Ishiwata H, Miyata Y, Kitaura R, et al. Evidence of diamond nanowires formed inside carbon nanotubes from diamantane dicarboxylic acid. Angewandte Chemie, International Edition. 2013;52:3717-3721. DOI: 10.1002/anie.201209192

63. Shellaiah M, Chen TH, Simon T, Li L-C, Sun KW, Ko F-H. An affordable wet chemical route to grow conducting hybrid graphite-diamond nanowires: Demonstration by a single nanowire device. Scientific Reports. 2017;7:11243. DOI: 10.1038/s41598-017-11741-9

64. Berman D, Deshmukh SA, Narayanan B, Sankaranarayanan SKRS, Yan Z, Balandin AA, et al. Metal-induced rapid transformation of diamond into single and multilayer graphene on wafer scale. Nature Communications. 2016;7:12099. DOI: 10.1038/ncomms12099

65. Barnard AS, Russo SP, Snook IK. Ab initio modeling of diamond nanowire structures. Nano Letters. 2003;3:1323-1328. DOI: 10.1021/nl034169x

66. Tanskanen JT, Linnolahti M, Karttunen AJ, Pakkanen TA. From fulleranes and icosahedral diamondoids to polyicosahedral nanowires: Structural, electronic, and mechanical characteristics. Journal of Physical Chemistry C. 2008;112:11122-11129. DOI: 10.1021/jp7119262

67. Barnard AS, Russo SP, Snook IK. From nanodiamond to diamond nanowires: Structural properties affected by dimension. Philosophical Magazine. 2004;84:899-907. DOI: 10.1080/14786430310001627412

68. Guo J, Wen B, Melnik R, Yao S, Li T. Molecular dynamics (Pembroke, Ont.) study on diamond nanowires mechanical properties: Strain rate, temperature and size dependent effects. Diamond and Related Materials. 2011;20:551-555. DOI: 10.1016/j.Diamond2011.02.016

69. Xue J, Jijun Z, Xin J. Mechanical and electronic properties of diamond nanowires under tensile strain from first principles. Nanotechnology. 2011;22:405705. DOI: 10.1088/0957-4484/22/40/405705

70. Trejo A, Miranda A, Niño de Rivera L, Díaz-Méndez A, Cruz-Irisson M. Phonon optical modes and electronic properties in Diamond nanowires. Microelectronic Engineering. 2012;90:92-95. DOI: 10.1016/j.mee.2011.04.052

71. Sun KW, Wang JY, Ko TY. Raman spectroscopy of single nanodiamond: Phonon-confinement effects. Applied Physics Letters. 2008;92:153115. DOI: 10.1063/1.2912029

72. Novikov NV, Podoba AP, Shmegera SV, Witek A, Zaitsev AM, Denisenko AB, et al. Influence of isotopic content on diamond thermal conductivity. Diamond and Related Materials. 1999;8:1602-1606. DOI: 10.1016/S0925-9635(99)00040-0

73. Moreland JF. The disparate thermal conductivity of carbon nanotubes and diamond nanowires studied by atomistic simulation. Microscale Thermophysical Engineering. 2004;8:61-69. DOI: 10.1080/10893950490272939

74. Guo J, Wen B, Melnik R, Yao S, Li T. Geometry and temperature dependent thermal conductivity of diamond nanowires: A non-equilibrium molecular dynamics (Pembroke, Ont.) study. Physica E: Low-dimensional Systems and Nanostructures. 2010;43:155-160. DOI: 10.1016/j.physe.2010.06.032

75. McCreery RL. Advanced carbon electrode materials for molecular electrochemistry. Chemical Reviews. 2008;108:2646-2687. DOI: 10.1021/cr068076m

76. Lv M, Wei M, Rong F, Terashima C, Fujishima A, Gu ZZ. Electrochemical detection of catechol based on As-grown and nanograss array boron-doped diamond electrodes. Electroanalysis. 2010;22:199-203. DOI: 10.1002/elan.200900296

77. Yang Y, Oh J-W, Kim Y-R, Terashima C, Fujishima A, Kim JS, et al. Enhanced electrogenerated chemiluminescence of a ruthenium tris(2,2[prime or minute]) bipyridyl/tripropylamine system on a boron-doped diamond nanograss array. Chemical Communications. 2010;46:5793-5795. DOI: 10.1039/C0CC00773K

78. Sankaran KJ, Kunuku S, Leou K-C, Tai N-H, Lin IN. Enhancement of the electron field emission properties of ultrananocrystalline diamond films via hydrogen post-treatment. ACS Applied Materials & Interfaces. 2014;6:14543-14551. DOI: 10.1021/am503823n

79. Saravanan A, Huang B-R, Sankaran KJ, Tai N-H, Lin IN. Highly conductive diamond–graphite nanohybrid films with enhanced electron field emission and microplasma illumination properties. ACS Applied Materials & Interfaces. 2015;7:14035-14042. DOI: 10.1021/acsami.5b03166

80. Lee JS, Liu KS, Lin IN. Electron field emission characteristics of planar diamond film array synthesized by chemical vapor deposition process. Applied Physics Letters. 1997;71:554-556. DOI: 10.1063/1.119607

81. Sankaran KJ, Lin Y-F, Jian W-B, Chen H-C, Panda K, Sundaravel B, et al. Structural and electrical properties of conducting diamond nanowires. ACS Applied Materials & Interfaces. 2013;5:1294-1301. DOI: 10.1021/am302430p

82. Sankaran Kamatchi J, Afsal M, Lou SC, Chen HC, Chen C, Lee CY, et al. Electron field emission enhancement of vertically aligned ultrananocrystalline diamond-coated ZnO core-shell heterostructured nanorods. Small. 2013;10:179-185. DOI: 10.1002/smll.201301293

83. Coffinier Y, Szunerits S, Drobecq H, Melnyk O, Boukherroub R. Diamond nanowires for highly sensitive matrix-free mass spectrometry analysis of small molecules. Nanoscale. 2012;4:231-238. DOI: 10.1039/C1NR11274K

84. Adamschik M, Kusterer J, Schmid P, Schad KB, Grobe D, Flöter A, et al. Diamond microwave micro relay. Diamond and Related Materials. 2002;11:672-676. DOI: 10.1016/S0925-9635(01)00619-7

85. Adiga VP, Sumant AV, Suresh S, Gudeman C, Auciello O, Carlisle JA, et al. Mechanical stiffness and dissipation in ultrananocrystalline diamond microresonators. Physical Review B. 2009;79:245403. DOI: 10.1103/PhysRevB.79.245403

86. Yang N, Uetsuka H, Williams Oliver A, Osawa E, Tokuda N, Nebel Christoph E. Vertically aligned diamond nanowires: Fabrication, characterization, and application for DNA sensing. Physica Status Solidi A. 2009;206:2048-2056. DOI: 10.1002/pssa.200982222

87. Nebel CE, Yang N, Uetsuka H, Osawa E, Tokuda N, Williams O. Diamond Nanowires, a new approach towards next generation electrochemical gene sensor platforms. Diamond and Related Materials. 2009;18:910-917. DOI: 10.1016/j.Diamond2008.11.024

88. Szunerits S, Coffinier Y, Galopin E, Brenner J, Boukherroub R. Preparation of boron-doped diamond nanowires and their application for sensitive electrochemical detection of tryptophan. Electrochemistry Communications. 2010;12:438-441. DOI: 10.1016/j.elecom.2010.01.014

89. Shalini J, Sankaran KJ, Dong C-L, Lee C-Y, Tai N-H, Lin IN. In situ detection of dopamine using nitrogen incorporated diamond nanowire electrode. Nanoscale. 2013;5:1159-1167. DOI: 10.1039/C2NR32939E

90. Peng X, Chu J, Wang L, Duan S, Feng P. Boron-doped diamond nanowires for CO gas sensing application. Sensors and Actuators B: Chemical. 2017;241:383-389. DOI: 10.1016/j.SNB2016.10.009

91. Tao Y, Degen CL. Single-crystal diamond nanowire tips for ultrasensitive force microscopy. Nano Letters. 2015;15:7893-7897. DOI: 10.1021/acs.nanolett.5b02885

3

Analysis of Electrochemical and Structurally Enhanced Limn2O^4 Nanowire Cathode System

1. Introduction

The energy storage field faces a critical challenge: namely, the development of rechargeable systems for load leveling applications (e.g. storing solar and wind energy). Among the available battery technologies to date, only Li-ion batteries may possess the power and energy densities necessary for high power applications. The Li ion battery interface materials can store a lot of Li ions but have large structure change and volume expansion, which can cause mechanical failure. In this work we exploited the use of nanowire (NW) cathode morphology to alleviate these issues. Nanowires offer advantages of a large surface to volume ratio, efficient electron conducting pathways and facile strain relaxation [1]. In lithium-ion batteries, the cathode plays a critical role in determining energy density. Here the main requirements are a prolonged cycle life, components (i.e., relevant elements) abundant in high quantities in the earth's crust, and environmentally friendly systems [2, 3]. Among the commonly used layered Ni or Co oxide materials, the spinel LiMn2O4 appears to be a more favorable cathode in lithium ion batteries [4]. Spinel LiMn2O4 is economical, nontoxic and a highly abundant material with superior safety [5]. Conversely, its common drawback is a kinetic limitation, which is observed under fast scan rate or high current density, when the characteristic two peaks/plateaus associated with the charge and discharge mechanism of the spinel structure diminishes. To overcome this obstacle and permit the use of LiMn2O4 in energy-demanding applications, the use of nanostructured morphologies for the development of fast kinetic electrodes is an ideal approach [6]. Literature studies have shown that the one-dimensional nanosized materials have faster kinetics and higher rate capability than micrometer-sized materials due to the large surface-to-volume ratio that enhances the contact between active material grains and electrolyte. However, the high- temperature sintering process, which is necessary for high- performance cathodes based on high-quality crystallinity, such as $LiMn2O4$ leads to large grain size and aggregation which alters the battery performance due to increased lithium ion diffusion length and decreased effective surface area contact with electrolyte. Here, the objective was to produce a highly crystalline nanostructured cathode electrode material. Single crystalline nanowire morphology has proven most appealing because the untwined material fabricated by the single crystalline nanowire reduces aggregation, electronic resistance and grain growth at elevated temperature [7]. Generally, the electrochemical performances of electrode materials are strongly influenced by the phase crystallinity, purity, particle size, and distribution. The internal channels in these nano-crystalline cathode material spheres serve two purposes. They admit liquid electrolyte to allow rapid entry of Li+ ions for quick battery charging, and they provide space to accommodate expansion and contraction during Li+ intercalation and deintercalation, boosting battery power characteristics critical to improve the LiMn2O4 performance. This research work produced highly crystalline LiMn2O4 nanowires, synthesized using a facile, easy to scale up process, starting with the preparation of R-MnO2 nanowires followed by solid state reaction with LiOH. Concomitantly, LiMn2O4 powders were also prepared and studied as comparison [8]. The high rate capability as well as phase stability of the nanowires architecture and electrochemistry was demonstrated as probed by electrochemical and spectrochemical characterization techniques. Because determination of the local structure seems one of the key issue for understanding electrochemical properties, vibrational spectroscopy was also applied to provide information on structural features of the nanowires.

2. Experimental procedure

2.1. Modified LiMn2O4 nanowire syntheses

In this work, the spinel LiMn2O4 powders were prepared following a procedure found in the literature with minor adjustments [9]. Typical synthesis includes the reaction of lithium hydroxide and manganese acetate (LiOH and Mn (CH3COO)2) via a co-precipitation method. A stoichiometric amount of LiOH and Mn(CH3COO)2 with the cationic ratio of Li/Mn = 1:2 were dissolved in deionized water and mixed by stirring. The solution is then evaporated at 100°C for 10 h to obtain the precursor powder [10]. Concomitantly we effectively produced ultrathin spinel LiMn2O4 nanowires using a facile, two-step process. First, single crystals were produced from a nonaqueous solution in an autoclave reaction to prepare R-MnO2 nanowires, followed by solid state reaction with LiOH. In a typical process, a hydroalcoholic solution was formed in distilled water and adding first (NH4)2SO4 with (NH4)2S2O8 and then 1-octanol. The solvothermal reaction was then performed at 140°C for 12 h in an autoclave to obtain α-MnO2. This was followed by a solid state reaction between R-MnO2 nanowires and LiOH at low pressure and oxygen atmosphere to achieve the pure LiMn2O4 nanowire phase. The chemistry of the nanostructures, the crystallinity and phase purity of LiMn2O4 powders, R-MnO2 and LiMn2O4 nanowires were all characterized by X-ray diffraction (XRD), scanning electron microscopy (FE-SEM), high resolution transmission electron microscopy (HR-TEM), electrochemical impedance spectroscopy and Nuclear magnetic resonance spectroscopy (7Li NMR) technique to observe the local magnetic fields around atomic nuclei.

3. Characterization and analysis

The morphology and particle size distribution of LiMn2O4 nanowire material was analyzed with a Transmission electron microscopy (TEM), taken with a JSM-6700F (JEOL). Raman spectra of all samples were recorded at wavelength of 514.5 nm using a laser-spectrometer (model Jobin-Yvon U1000) equipped with dual monochromators coupled with special filter and emission of an argon-ion laser (model Spectra-Physics 2020). To avoid sample photodecomposition or denaturation a power density as low as 10 W cm−2 was used. NMR was carried out at room temperature on Varian spectrometer (VNMRS WB 500 solids) with an 11.7 T magnet. Magic angle spinning (MAS) NMR experiments were performed using a 6 mm MAS probe. The 7Li resonance frequency was 194.29 MHz with a sample spinning speed of 16 kHz. Single-pulse and echo-pulse sequences were used to acquire all spectra. The amount of lithium detectable in the materials was quantified using 1 M LiCl (116.571 MHz) as an external reference for chemical. XRD Measurements were carried out with a D8 ADVANCE diffractometer from BRUKER axs using an X-ray tube with copper K-alpha radiation operated at 40 kV and 40 mA and a position sensitive detector, Vantec_1, which enables fast data acquisition. Measurement range: [12–90° in 2 theta], Step size: 0.027° in 2 theta, Measurement time: 1 s/step.

3.1. Electrochemical tests

The Electrochemistry of the LiMn2O4 nanowire cathode, were carried out with electrodes mounted on coin-cells operating at 30°C. Electrode mixtures were prepared by mixing the oxide powder (70 wt%) with acetylene black (current collector) (20 wt%) and polyvinylidene fluoride (PVdF) binder (10 wt%) in N-methylpyrrolidone (NMP) solvent to form a mixed slurry. The slurry was coated on an aluminum foil, followed by drying in a vacuum oven at 120°C for 48 h and a cathode electrode was formed.

4. Results and discussions

4.1. Surface morphology

4.1.1. Transmission electron microscopy (TEM)

Figure 1a–d shows the TEM micrographs of the crystalline LiMn2O4 spinel nanowires in the range of 5–20 nm without agglomeration. The wires appear to be highly crystalline and moderately dispersed, which causes the material to possess a larger surface area. These nanowires have diameters of tens of nanometers and lengths up to several micrometers and basically retains the morphology of the precursor MnO2 nanowires. At higher magnification, Figure 1c, nanowires adhering together are observed, which leads to the notion that these nanowires have a fiber-brush aspect. Figure 1b is a typical high-resolution electron microscopy (HREM) image clearly displaying the lattice fringes of the material. The SAED patterns along [100] direction of the single crystals are showed in the inset. Bright diffraction spots including (022), (222) and (004) in the [100] zone are generated from the spinel structure with Fd3m space group. The images also show single LiMn2O4 nanowires with distinctive lattice fringes. The discriminable lattice fringes illustrate that the prepared nanowires are single crystals in the area shown. Figure 2 shows a typical TEM image at a magnification of 2 and 20 nm of LiMn2O4. The primary particle size of the as-synthesized powders is around 10 nm with visible agglomerations. The as-synthesized and calcined powders typically have surface areas of 18.0 m2 g−1 measured by BET method [11].

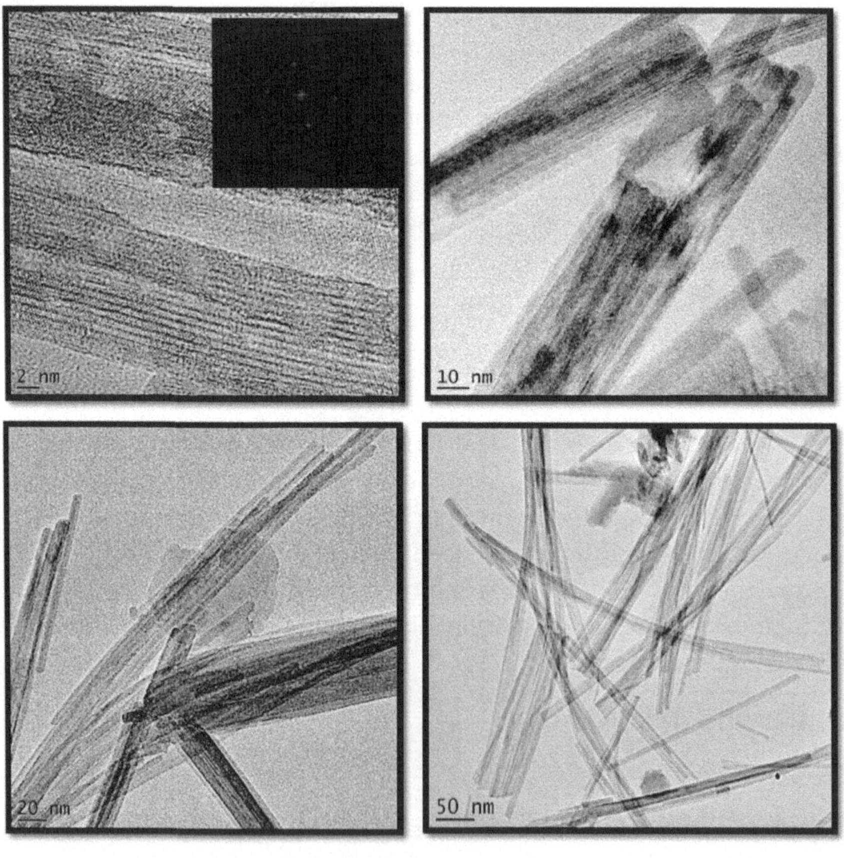

Figure 1. High resolution TEM image of crystalline LiMn2O4 nanowires.

Figure 2. TEM images of LiMn2O4 nanoparticles.

4.1.2. Atomic force microscopy (AFM)

The surfaces of the cathode were observed by high-resolution atomic force microscopy in dried state. Figure 3 shows 2 × 2 μm2 dimensional AFM images of LiMn2O4 nanowires and the LiMn2O4 nanopowders in Ar atmosphere. The images reveal apparent changes in roughness between the films. It is known that the distribution of particles influences the cyclability and discharge capacity [12]. Tapping mode AFM images displayed the surface morphology of nano LiMn2O4 particles (a–b). Evidently, a rough structure with closely distributed micro-pores of less than 5 nm in diameter was observed. From the two-dimensional image, it was evident that this surface yields a large degree of surface roughness (a). A more detailed analysis of the particle size is shown in the histogram of (a, b-i). Particle sizing and metrics play a critical role in determining battery capacity and performance. Therefore, the typical size of materials used for battery construction is >1 μm [13]. Here, the size distribution moved toward an average diameter of 60 nm aiding the high-rate capabilities of the cathode. For LiMn2O4 several "cauliflower-like" areas are observed with a rms (root mean square) roughness of 73.5 Å, whilst LiMn2O4 nanowires topography appears smoother and homogeneous with rms roughness of 26.3 Å which can be calculated using Eq. (1). These features constitute the stability of the spinel structure, which enhances the electrochemical properties.

$$R_q = \sqrt{\frac{1}{n}\sum_{i=1}^{n} y_i^2} \tag{1}$$

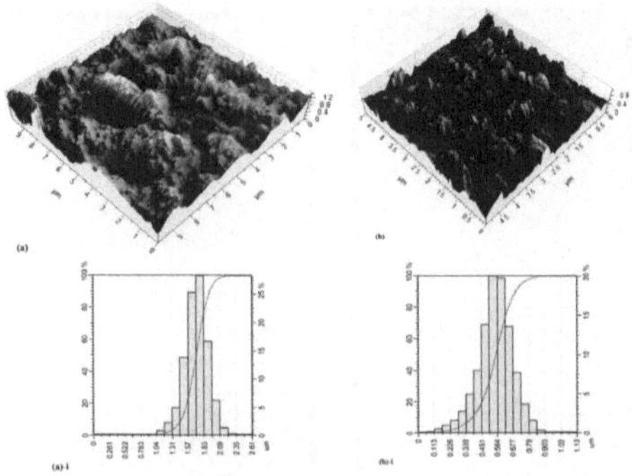

Figure 3. AFM images of LiMn2O4 shown in tapping mode in three-dimensional views at 600 nm and particle size distribution histogram (i) of LiMnO4 (a) and LiMn2O4 nanowires (b), respectively.

4.1.3. Vibrational structure analysis (Raman/SS-NMR)

Figure 4 shows the Raman spectra (RS) of LiMn2O4 and LiMn2O4 nanowires. A Blue shift of photons and a noticeable decrease in the peak width is observed with modification. The bands at ~560 and ~660 cm−1 are attributed to the O-Mn-O bending and stretching modes, respectively. Some vibrational or rotational transitions, which exhibit low polarizability, becomes Raman inactive therefore sharper peaks signifies better crystallinity and less cation mixing [14]. The observation of narrower bands and less modes may be the result of no translation invariance and lattice distortion around the Mn3+ and Mn4+ cations. The catalytically active tetragonal hausmannite (Mn3O4) spinel is acknowledged by the solid peak erected at 633 cm−1, conforming to the Mn-O breathing vibration of Mn2+ ions in tetrahedral coordination (A1g mode) [15]. The T2 g (1), Eg, and T2 g (2) modes of Mn3O4 is shown by peaks 300, 347, and 492 cm−1 respectively. As discussed by Julien et al. [16], the A1g mode correlated with Mn-O vibration of MnO6 groups will shift to lower energies as the average Mn oxidation state increases. The Raman data are in agreement with diffraction analysis.

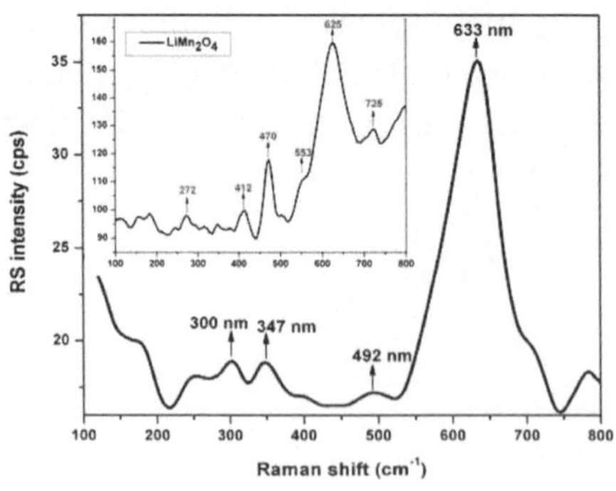

Figure 4. Vibrational spectra of LiMn2O4 nanowires and LiMn2O4 precursor (inset).

Figure 5 shows the 7Li-NMR spectrums of both spinel LiMn2O4 and LiMn2O4 nanowires.

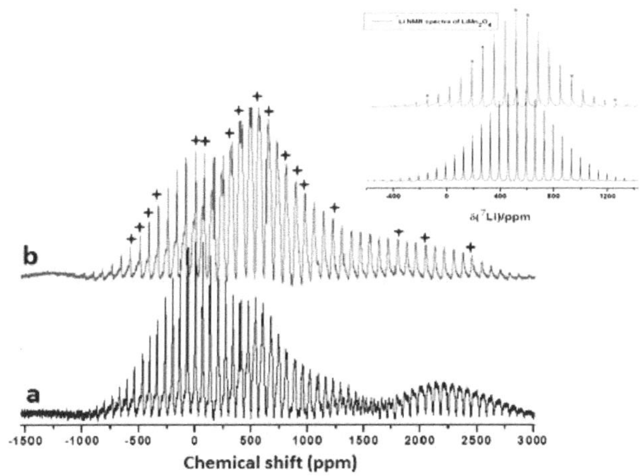

Figure 5. LiMn2O4 nanowire NMR spectrum at 13 kHz (a) and 16 kHz (b) spinning speed, magnetic field strength of 11.7 T and resonance frequency of 194.29 MHz. LiMn2O4 nanopowder (inset).

Nuclear magnetic resonance (NMR) spectroscopy has been employed as a significant tool to probe the local structure and dynamics of these nano materials. Here we show how this technique help understand the origins of the performance of the given nanomaterial [17]. LiMn2O4 is a hopping semiconductor containing both Mn3+ and Mn4+ ions. The hopping time scale is fast between these ions in comparison to the NMR time scale (ca. 10−5 s) and therefore the lithium spins detect a manganese oxidation state corresponding to 3.5 (i.e., "Mn3.5+" ions), pertaining to only one magnetically in equivalent lithium site (the 8a site) [12]. This permits the NMR spectroscopy to be used as a means to follow the partial charge-ordering process. The hyperfine shift of >500 ppm from the chemical shift range of diamagnetic compounds containing lithium is characteristic of the 7Li MAS NMR spectrum of LiMn2O4.

The spinel LiMn2O4 nanowires have a distinctly different spectral line shape as compared to their powder form [18]. The isotropic resonance at 511 ppm for LiMnO4 is assigned to lithium ions in the tetrahedral 8a site, whilst the isotropic resonances at ~680 and 835 ppm for LiMn2O4 nanowires is assigned to lithium present in the proximity of higher oxidation state manganese ions (Mn4+). The latter is generally originated from vacancies on both the lithium and manganese sites (i.e., Mn 16d sites and the interstitial 16c sites or Li-for-Mn substitutions) [19]. The resonance at 445 ppm is ascribed to the presence of Mn3+ ions in the Li local coordination sphere. Mn3+ is a Jahn–Teller (distortion) active ion; therefore there will be a distortion of the octahedron in this case [20]. The enhancement of spinning sideband manifold for LiMn2O4 nanowires is caused by the increased portion of paramagnetic manganese around lithium [21]. Hence, it can be suggested that the lithium atoms are interacted with the manganese and cause for better electrochemical performance.

4.1.4. X-ray diffraction microscopic analysis (XRD)

The lattice constant was calculated from the corresponding diffraction pattern using XRD spectra, in relation to the crystal structure and is reported in Figure 6. All diffraction peaks can be assigned to the diffraction indices of LiMn2O4 spinel (JCPDS file no. 35-782), indicating that the structure of the spinel was maintained. The majority diffraction peaks of LiMnO2 are observed, and closely correspond to layered LiMnO2 (011), (202) and (111) planes. The intensive diffraction peaks appeared at 21.92, 36.92, 42.42 and 55.96°, respectively, should be assigned to the characteristic peaks for γ-MnO2, and the peaks occurred at 17.94, 28.78, 66.0°, respectively, should be ascribed to the characteristic peaks for α-MnO2. Hence, the sample appears to be composed of both γ and α-MnO2. From the broadening of XRD peaks, it can be said that nanowires are being formed. This suggest that the nanowires would render unchallenging Li+ pass through the coating layer more during charging and discharging process as the diffusion path is unhindered [22, 23]. The strong peak at $2\theta = 18.76°$, corresponds to a (111) peak with an interplanar distance of d = 0.476 nm. The full width at half-maximum (FWHM) becoming narrower is also due to higher synthesis temperatures which help to enhance the mobility of atoms. Sharp and relatively high intensity peaks (as compared to LiMn2O4) depicts high crystallinity [24].

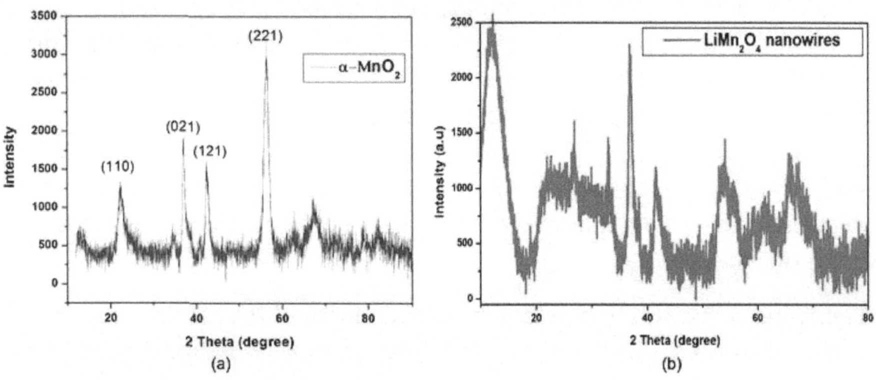

Figure 6. XRD spectra of LiMn2O4 nanowires (a) and alfa-MnO2 (b).

4.2. Electrochemical analysis

4.2.1. Redox reaction analysis

To clarify the kinetic behavior of lithium-ion transfer, the discharge tests were carried out at. Figure 7 shows the effect of discharge current densities on the capacities of the two types of electrodes. It is well known that the nanowire electrodes have good electronic conductivity; therefore, they can greatly increase the electrical conductivity among the transition metal oxide particles and reduce the polarization of the LiMn2O4 electrode. The nanowire cathode shows an increased discharge current densities due to improvement of Li+ diffusion pathway [25, 26, 27]. This coincides with prior research that has recognized the ability of nanowire architecture to enhance the stability or cyclability of electrode materials [28]. The initial discharge curve of LiMn2O4 nanowire electrode was similar to that of LiMn2O4 spinel, due to the lower ratio of Mn3+/Mn4+ ions [29]. The initial discharge capacities and corresponding coulombic efficiencies after cycling for the nanowires and LiMn2O4 were 146 mAh g−1 / 99% and 122 mAh g−1/ 70%, respectively. The decrease in capacity over subsequent cycles is explained by the change in surface area [30]. Three-dimensional porous nanostructures with large surface area could exhibit higher durability in the lithium insertion/ extraction process at a high current density, owing to the short lithium ion diffusion lengths in the 3-D channels of the electrode [31]. The LiMn2O4 nanowire electrode exhibited excellent rate capability as shown in Figure 7. The large capacities are due to the nanowire morphology, stability and the high quality of the single crystal, which can shorten the diffusion lengths of both the lithium and electrons [32]. This was further corroborated by electrochemical impedance spectroscopy results shown in Figure 8. The nanowires show a significant decrease in impedance due to their enhanced electrochemical diffusion processes.

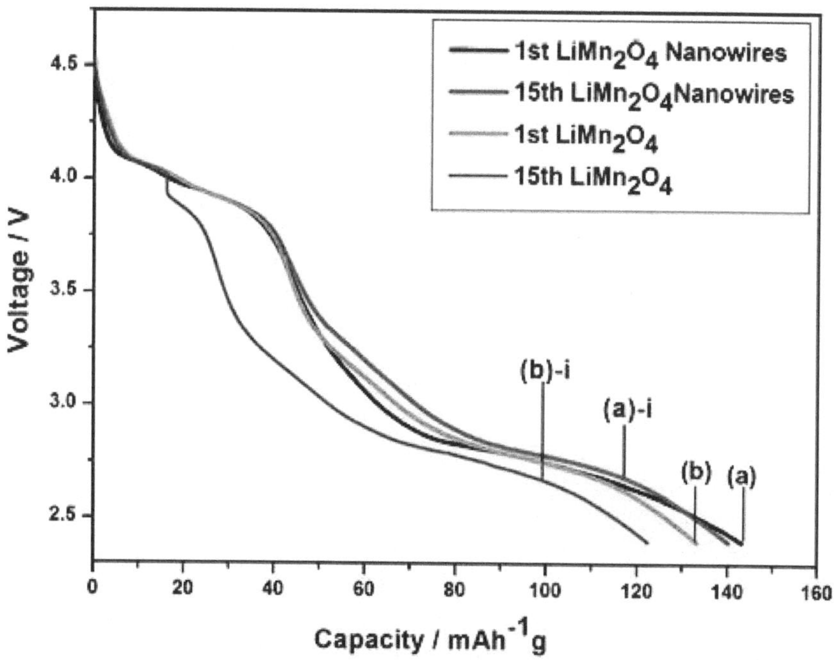

Figure 7. Discharge current densities at 0.1 mV s−1 for LiMn2O4 nanowires (a) and LiMn2O4 (b) in EC: DMC, 1 M LiPF6 for the (1st (a, b) and 50th (a-i, b-i) cycle).

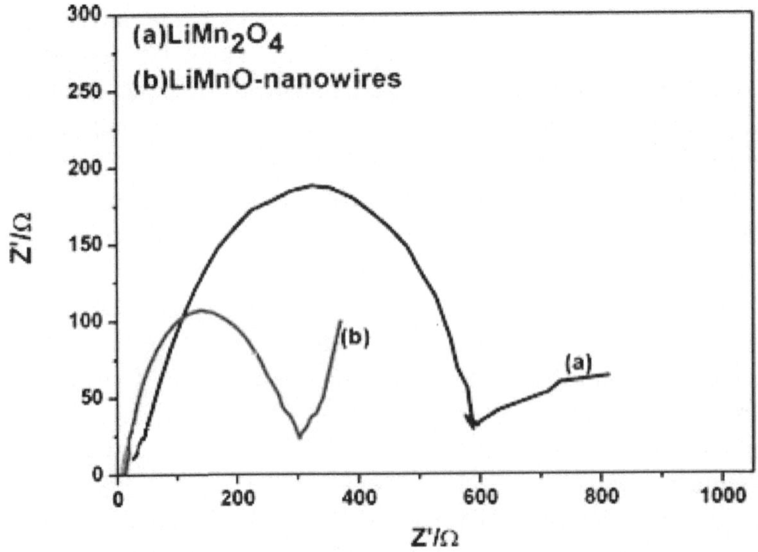

Figure 8. Nyquist plot of LiMn2O4 (a) and LiMn2O4 nanowires (b).

5. Conclusion

In summary, the LiMn2O4 nanowires have proven excellent thermal stability for a high-temperature sintering process as well as a charge-discharge reversible stability and improved conductivity attained by their architecture, excellent crystallinity and decreased impedance. All the as-prepared powders were identified as a single phase of cubic spinel structure indicative of an unobstructed Li ion diffusion pathway. The results have shown that thin-nanowire precursor morphology is preserved after the solid-state reaction. Such morphology improves the kinetic properties at very high current rate and was capable of the facile structural transformation of the cubic and tetragonal phase in the large compositional range. The LiMn2O4 nanowires showed a decrease in potential difference, indicating an improved charge transportation process. The nanostructures aid structural stability, reduction of side reactions and Mn dissolution between the interface of the cathode and electrolyte, which contributes to the recovering performance. Moreover, the nanowire cathode system has great potential for improving the electrode-filled ratio and safety in lithium ion battery–operating electronic devices, in transportation applications, and in applications on the electric grid.

References

1. Marom R, Amalraj SF, Leifer N, Jacob D, Aurbach D. A review of advanced and practical lithium battery materials. Journal of Materials Chemistry. 2011;21:9938
2. Kim YS, Kanoh H, Horotsu T, Ooi K. Chemical bonding of ion-exchange type sites in spinel-type manganese oxides Li1.33Mn1.67O4. Materials Research Bulletin. 2002;37:391-396

3. Thackeray MM, David WIF, Goodenough JB. Lithium insertion into manganese spinels. Materials Research Bulletin. 1983;18:641-647
4. Zhong Q, Bonaldarpour A, Zhang M, Gao Y, Dahn JR. Synthesis and electrochemistry of LiNixMn2 – xO4. Journal of the Electrochemical Society. 1997;144:205
5. Dokyun K, Hyun-Wook L, Muralidharan P, Cui Y. Spinel LiMn2O4 nanorods as lithium ion battery cathodes. Nano Letters. 2008;8(11):3948-3952
6. Sun YK, Jeon YS, Leeb HJ. Overcoming Jahn-Teller distortion for spinel Mn phase. Electro-chemical and Solid-State Letters. 2000;3:7
7. Hosono E, Kudo T, Honma I, Matsuda H, Zhou H. Synthesis of single crystalline spinel LiMn2O4 nanowires for a lithium ion battery with high power density. Nano Letters. 2009;9(3):1045-1051
8. Hyun-Wook L, Muralidharan P, Riccardo R, Claudio MM, Yi C, Kyung KD. Ultrathin spinel LiMnO nanowires as high power cathode materials for Li-ion batteries. Nano Letters. 2010;10(10):3852-3856
9. Hui X, Zhentao L, Jianping X. Nanostructured LiMn2O4 and their composites as high-performance cathodes for lithium-ion batteries. Progress in Natural Science: Materials International. 2012;22(6):572-584
10. Kweon HJ, Kim SJ, Park DG. Modification of LixNi1-yCoyO2 by applying a surface coating of MgO. Journal of Power Sources. 2000;88:255-261
11. Xiuqiang X, Dawei S, Bing S, Jinqiang Z, Chengyin W, Guoxiu W. Synthesis of single-crystalline spinel LiMn2O4 nanorods for lithium-ion batteries with high rate capability and long cycle life. Chemistry A european Journal. 2014;20(51):17125-17131
12. Wang P, Li YN, Yang J, Zheng Y. Carbon-coated Si-Cu/graphite composite as anode material for lithium-ion batteries. International Journal of Electrochemical Science. 2006;1:122-129
13. Li W, Andrei D, Pilgun O, Hugo C, Suhyeon P, Jaephil C, Arumugam M. Dynamic behaviour of interphases and its implication on high-energy-density cathode materials in lithium-ion batteries. Nature Communications. 2017;8:14589
14. Ramana C, Massot VM, Julien CM. XPS and Raman spectroscopic characterization of LiMn2O4 spinels. Surface and Interface Analysis. 2005;34:412-416
15. Li TT, Guo CL, Sun B, Li T, Li YG, Hou LF, Wei YH. Ell shaped Mn3O4 tetragonal bipyramids with good performance for lithium ion batteries. Journal of Materials Chemistry A. 2015;3:7248-7254
16. Julien CM, Massot M, Poinsignon C. Lattice vibrations of manganese oxides: Part I. Periodic structures. Spectrochimica Acta A. 2004;60:689-700
17. Jordi C, Clare PG. Lithium-ion batteries: Li-6 MAS NMR studies on materials. 2011. DOI: 10.1002/9781119951438.eibc0457
18. Clare PG, Nicolas D. NMR studies of cathode materials for lithium-ion rechargeable batteries. Chemical Reviews. 2004;104:4493-4512

19. Chi-Hung L, Chun-Wei H, Jui-Yuan C, Chung-Hua C, Tzung CT, Kuo-Chang L, Ming-Yen L, Wen-Wei W. Optoelectronic properties of single-crystalline Zn2GeO4 nanowires. Journal of Physical Chemistry C. 2014;118(15):8194-8199

20. Naghash AR, Lee JY. Preparation of spinel lithium manganese oxide by aqueous co-precipitation. Journal of Power Sources. 2000;85:284

21. Lee YJ, Wang F, Crey CP. 6Li and 7Li MAS NMR studies of lithium manganate cathode materials. Journal of the American Chemical Society. 1998;120:2601

22. Byung-Sik M, Jae-Hyeong L, Jung H. Comparative studies of the properties of CdS films deposited on different substrates by R.F. sputtering. Thin Solid Films. 2006;511:299-303

23. Hunter JC. Preparation of a new crystal form of manganese dioxide. Journal of Solid State Chemistry. 1981;39:142-147

24. Eftekhari A. Effects of metal source in metal substitution of lithium manganese oxide spinel. Solid State Communications. 2006;140:39

25. Ni JF, Zhou HH, Chen JT, Zhang XX. LiFePO4 doped with ions prepared by co-precipitation method. Materials Letters. 2005;59:2356-2361

26. Yu L, Cai D, Wang H, Titirici M-M. Synthesis of microspherical LiFePO4-carbon composites for lithium-ion batteries. Nanomaterials. 2013;3:443-452

27. Atanasov M, Barras J-L, Benco L, Daul C. Electronic structure, chemical bonding, and vibronic coupling in MnIV/MnIII mixed valent LixMn2O4 spinels and their effect on the dynamics of intercalated Li: A cluster study using DFT. Journal of the American Chemical Society. 2000;122:4718-4728

28. Yi T-F, Li C-Y, Zhu Y-R, Zhu R-S, Shu J. Electrochemical intercalation kinetics of lithium ions for spinel LiNi0.5Mn1.5O4 cathode material. Russian Journal of Electrochemistry. 2010;46:227-232

29. Yamada A, Chung S, Hinokuma CK. Optimized LiFePO4 for lithium battery cathodes. Journal of the Electrochemical Society. 2001;148:A224

30. Yin Z, Ma D, Bao XH. Emulsion-assisted synthesis of monodisperse binary metal nano-particles. Chemical Communications. 2010;46:1344-1346

31. Kim JM, Lee G, Kim BH, Huh YS, Lee G-W, Kim HJ. Ultrasound assisted synthesis of Li-rich mesoporous LiMn2O4 nano-spheres for enhancing the electrochemical performance in Li-ion secondary batteries. Ultrasonics Sonochemistry. 2012;19:627-631

32. Qu Q, Fu L, Zhan X, Samuelis D, Maier J, Li L, Tian S, Li Z, Wu Y. Porous LiMn2O4 as cathode material with high power and excellent cycling for aqueous rechargeable lithium batteries. Energy & Environmental Science. 2011;4:3985-3990

4

ZnO Nanowires Growth Direction and Parameters Affecting Their Surface Morphology

1. Introduction

Recently, nanomaterials have attracted much attention due to their high performance in electronics, optics, and photonics. Nanomaterials can be classified into three groups—zero dimensional, one dimensional, and two dimensional—due to their aspect ratio. Zero-dimensional nanostructures are referred to as quantum dots or nanoparticles with an aspect ratio near unity. One-dimensional nanostructures such as nanowires, nanorods, nanofibers, nanobelts, and nanotubes find perfect application in semiconductor materials and device. Two-dimensional nanomaterials, as thin films, are used as an optical coating, corrosion protection, and semiconductor thin film devices.

Currently, one-dimensional zinc oxide (ZnO) nanowire is one of the most interesting nanomaterials for researchers [1]. It is a semiconductor material with a direct wide band gap energy (3.37 eV) and a large excitation binding energy (60 meV) at room temperature [2]. ZnO exists in two main crystalline structure, hexagonal wurtzite and cubic zinc, blended as shown in Figure 1.

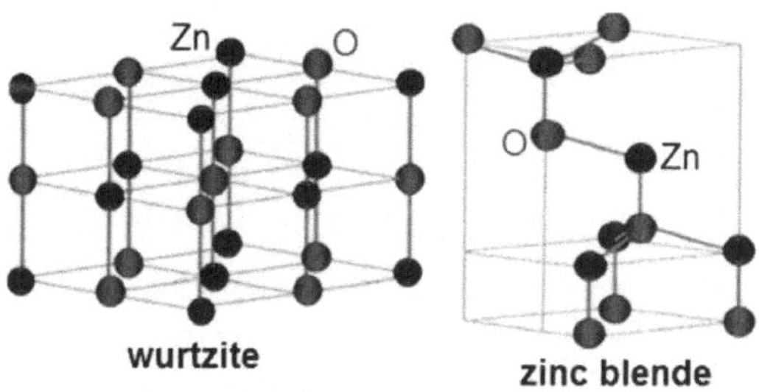

Figure 1. Two types of ZnO crystal structure.

The hexagonal crystalline structures with about 1.60 of c/a ratio is close to the hexagonal cell ideal c/a value = 1.633 [10]. Its structure could be described as a number of alternating planes composed of tetrahedrally coordinated of $O2-$ and $Zn2+$ stacked along the c-axis (Figure 1) [3].

Due to their remarkable performance in electronics, optics, and photonics, ZnO nanowires are attractive candidates for many applications such as UV lasers [4], light-emitting diodes [5], solar cells [6, 7], gas sensors [8], photodetectors [9], and photocatalysts [10]. ZnO nanowires also are used increasingly as photocatalysts to inactivate bacteria and viruses and for the degradation of environmental pollutants such as dye pesticides and volatile organic compounds [11, 12].

2. ZnO nanowire synthesis

ZnO nanowires can be synthesized mainly by two methods: vapor phase and solution phase. Solution phase synthesis has many advantages over vapor phase synthesis, as low cost, low synthesis temperature (<200°C), scalability, and ease of handling. This allows using different kinds of substrates including inorganic and organic, an aqueous or organic solution, or a mixture of them in the growth process [13, 14].

3. Hydrothermal method

The solution phase synthesis processes carried out in an aqueous solution, under high vapor pressure and temperature [4, 15]. Hydrothermal method among all of the synthesis methods is found to be the most attractive synthesis technique to obtain well-aligned ZnO nanowires, hierarchical or branched nanowires. Since no need for high-temperature manufacturing and vacuum processing in hydrothermal method that would offer the potential for much lower processing cost. A lot of effort has been made to investigate the effect of synthesis parameters on the density, dimensions, and quality of ZnO nanowires [3, 4, 6, 7, 15, 16]. Many focused on seeding layer effect [10, 17, 18, 19, 20], chemical reagent, and their concentration effect on ZnO nanowire density [21], morphology [22], growth time, and temperature [22]. We think that, there are still need to study the effect of additives under different fabrication conditions on ZnO nanostructure morphology.

The general process for vertically aligned ZnO nanowires grown on a seeded or unseeded substrate by the hydrothermal method is done using mixture aqueous solution as a precursor (or growth solution) that contains an alkaline reagent and $Zn2+$ salt (Zn(NO3)2, ZnCl2, etc.). The alkaline reagents that have been used to supply $OH-$ during the reaction process are NaOH, hexamethylenetetramine (HMTA), Na2CO3, ammonia, and ethylenediamine. Elevated temperatures (>100°C) and pressures in a Teflon-sealed stainless autoclave [24, 25, 26, 27] are required when NaOH, KOH, or Na2CO3 is chosen to supply $OH-$ in the synthesis process.

HMTA, ammonia, or ethylenediamine has been chosen in the synthesis process which can reduce the synthesis temperatures (<100°C) and pressure to an atmospheric pressure [28].

It was found that the key parameters for controlling nucleation and morphology of ZnO nanowires can be achieved by controlling the supersaturation reaction and reactant stability that depends on solution temperature and pH, precursor concentration, and duration growth time. Another important parameter is the nature of the substrate on which ZnO nanowires nucleate and grow over. Seeding of ZnO particles is usually used to prompt the nucleation process and improve ZnO particles adhesion, their distribution, and alignment. Without a ZnO seeding layer, ZnO nanowires could be grown also on a seedless substrate by using Au and introducing a suitable content of ammonium hydroxide into the precursor solution. Au is used as an "intermediate layer" to assist the growth of ZnO nanowires after treating the layer [29].

4. ZnO nanowire synthesis mechanism

Hexamethylenetetramine ((CH2)6N4 or HTMA) and Zn(NO3)2 are the most often used in producing high-quality ZnO nanowires [28].

Nanowire formed after ZnO cluster can be precipitated on the substrate when the growth solution attained saturation when the production of anion and cation exceeds the solubility of the reactants [29].

Hexamethylenetetramine (HMT) was responsible for the production of ammonia (NH3) that played the role of complexion agent. That is, complexion agent aided the yield of Zinc tetramine (Zn(NH3)42+) complexion in alkaline medium. The reaction that occurs in the bath is shown in the chemical equations [30]:

Decomposition reaction:

$CH2\ 6\ N\ 4 + 6H\ 2\ O \rightarrow 6HCHO + 4NH\ 3$ E1

Hydroxyl supply reaction:

$NH\ 3 + H\ 2\ O \rightarrow NH\ 4 + + OH -$ E2

$Zn\ NO\ 3\ 2 \rightarrow Zn\ 2 + + 2NO\ 3 -$ E3

$NO\ 3 - + H\ 2\ O + 2e - \rightarrow NO\ 2 - + 2OH -$ E4

Supersaturation reaction:

$Zn\ 2 + + 2OH - \rightarrow Zn\ OH\ 4\ 2 -$ E5

$Zn\ 2 + + 4NH\ 3 \rightarrow Zn\ NH\ 3\ 4\ 2 +$ E6

ZnO nanowire growth reaction:

Zn NH 3 4 2 + + 2OH — → ZnO + 4NH 3 + H 2 O E7

Zn OH 4 2 — → ZnO cluster + H 2 O + 2OH – E8

In the growth solution, the hexamethylenetetramine is first hydrolyzed (Eq. (1)). NO3– ions from Zn (NO3)2 decomposition acquire electrons and are reduced to NO2– ions, inducing the increase of OH–concentration (Eq. (4)) [32]. OH– ions are combined with Zn2+ ions to form an intermediate growth unit of Zn(OH)42– (Eq. (5)), while Zinc tetramine (Zn(NH3)42+) is formed from the combination of Zn2+ ions with NH4+. Due to dehydration of the complex in the alkaline medium, ZnO clusters are formed by the electrostatic repulsion of ionic species (Zn2+ and OH–) (Eqs. (7) and (8)). It is well known that solution supersaturation is a prerequisite for ZnO growth and is also intimately connected with their growth and morphology evolution. As the reactions continue, more ZnO clusters appear in solution. When the solution becomes supersaturated, nucleation cluster of molecules formed undergo rapid decomposition, and particles combine to produce film thickness, where ZnO nanowire is formed in the solution via the substrate.

One of the key parameters for the growth of ZnO nanowires is controlling the reactant supersaturation. It is believed that high supersaturation levels favor nucleation and low supersaturation levels favor crystal growth. With a short explanation for this, they reported that if a lot of OH– is produced in a short period, the Zn2+ ions in the solution will precipitate out quickly, and therefore, Zn2+ would contribute little to the ZnO nanowire growth and result in the fast consumption of the nutrient and prohibit further growth of the ZnO nanowires. Thus, the concentration of OH– should be controlled in the solution to maintain s suitable per saturation levels during the whole nanowire growth process [32].

HMTA control growth direction due to its ability to attach the nonpolar facets of the ZnO nanowires and prevent attracting access Zn2+ ion, thus leaving only the polar (001) face for epitaxial growth [32].

5. Parameters affecting ZnO nanowire growth and morphology

5.1. Precursor concentration

Wang et al. reported that unless the average diameter and length of ZnO nanorods increases with the precursor concentration increases (Figure 2), it was found that the changes in the [Zn(NO3)2]/[C6H12N4] ratio have no significant effect on the diameters of the ZnO nanorods (Figure 3).

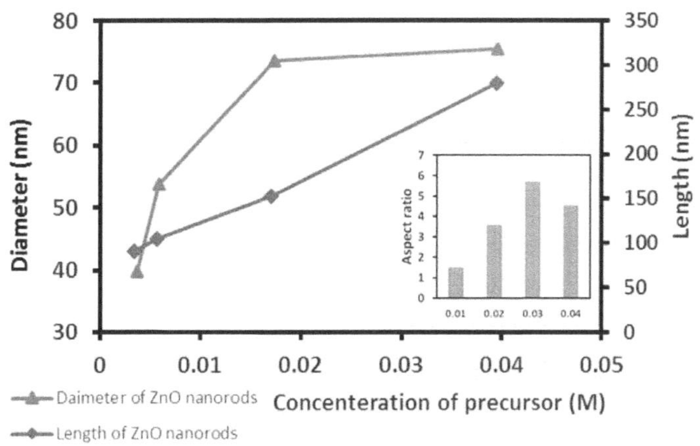

Figure 2. Average diameters, lengths, and aspect ratios of ZnO nanorods prepared from various precursor concentrations with a Zn(NO3)2/ C6H12N4 ratio of 5 (image from [33] ©2008 The American Ceramic Society).

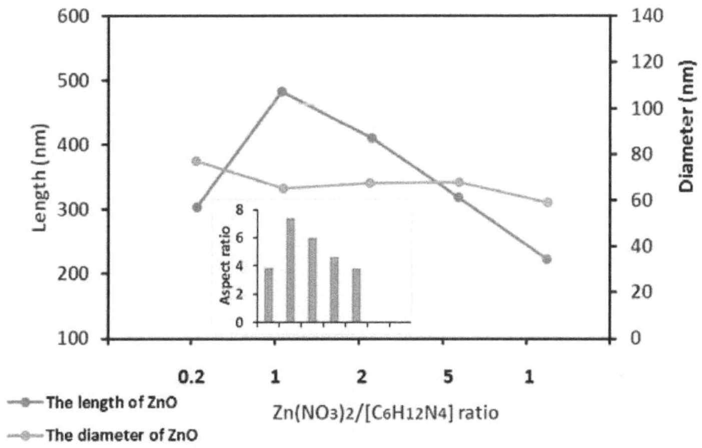

Figure 3. Average diameters, lengths, and aspect ratios of ZnO nanorods prepared in a 0.04 M precursor solution with various Zn(NO3)2/ C6H12N4 ratios (image from [33] ©2008 The American Ceramic Society).

It was found also that using unity ratio leads to achieve the maximum value of aspect ratio of ZnO nanorod array [33].

Xu et al. synthesized ZnO nanowires using equimolar concentrations of the zinc salt and HMTA; the nanowire density increased till reaching steady state with the precursor concentration. The authors referred that to the increase of zinc chemical potential with zinc concentration in the solution. To balance this increment, more nucleation sites on the substrate surface will be generated, and, therefore, the density of the ZnO nanowires will increase. However, when the density reaches saturation rate, any increase in the solution concentration may not increase the density of the nanowires anymore [8].

Kim et al. reported that precursor concentration plays a role in ZnO nanowire structure and increasing the concentration could change ZnO nanorod growth from single crystals with low density and variable orientations to polycrystals due to the supersaturated Zn2+ source [34].

Gokarna et al. achieved well-aligned, highly crystalline, small-diameter urchin-like ZnO NWs on PS beads which were pre-coated with a ZnO seed layer. They found that the average diameter of ultra-thin ZnO NWs that are produced from salt precursor dissolved in water with the small addition of NH4OH as growth solution could be controlled by varying the concentration of zinc acetate in the growth solution. Smaller size is obtained with lower concentration. XRD patterns showed that the NWs had a wurtzite hexagonal structure and (001) preferred growth direction [35] (Figure 4).

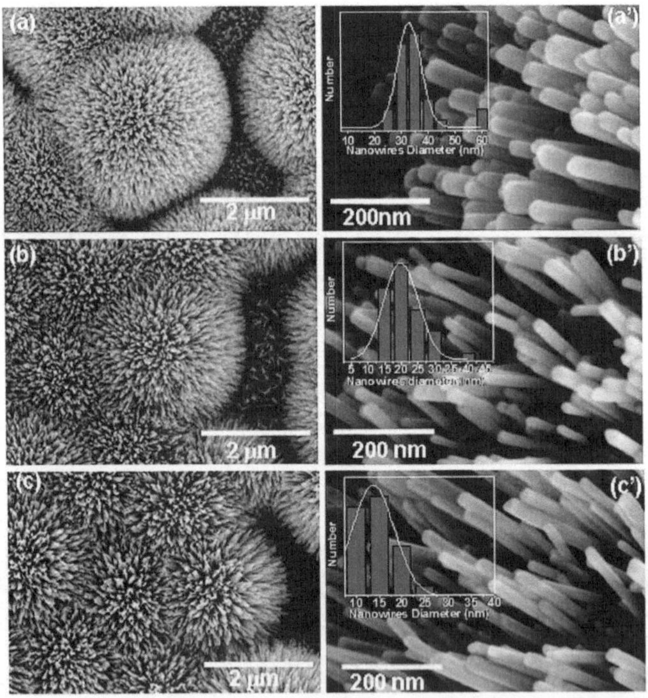

Figure 4. SEM images of ZnO NWs grown as a function of varying solution concentration. (a, a') 0.029 M, (b, b') 0.02 M, and (c, c') 0.018 M, respectively. (a, b, c) are at low magnification; (a', b', c') are at high magnification of the NWs with the diameter distribution of the NWs. The white line on the histograms represents the corresponding fitted Gaussian distributions (image from [35]).

In our previous work, we studied ZnO nanowire structure and morphology by varying the precursor concentration with equimolar concentrations of the zinc salt and HMTA on the seedless substrate. We found that increasing precursor concentration might provide a lot of ZnO nuclei and densify nanorods. For samples fabricated at low precursor equimolar solution (5 mM), nonalignment, small aspect ratio, low densities, and prism-like ZnO nanorods were achieved, Figure 5a. Increasing precursor concentration to 10 Mm leads to decreased aspect ratio and increased density of nonalignment prism-like-shaped nanorods, Figure 5b. Figure 5 shows the surface morphology of the above samples [37].

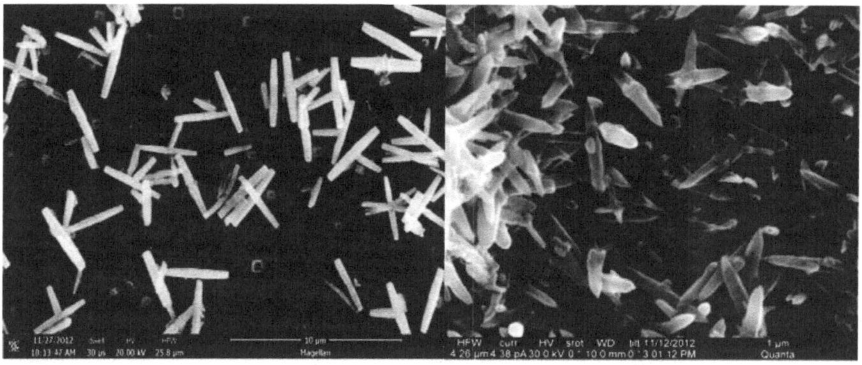

a b

Figure 5. SEM image of ZnO grown at precursor equimolar solution: (a) (5 mM) and (b) (10 mM) (image from [37]).

5.2. Growth temperature

ZnO nanowire growth by a hydrothermal process is a thermo-activated process; nanowire formation must be activated by the heating of growth solution to the desired temperature. A minimum temperature of ~60°C is required to obtain ZnO nanowires. Cesar and his group [38] reported that the starting temperature of ZnO growth solution on seeded substrate plays a role in their resultant morphologies. Bigger nonaligned ZnO nanowires with a slight coneshaped achieved when the growth solution heated from room temperature to the required temperature during synthesization process, while vertically aligned small-diameter ZnO nanowires achieved when pre-heated solution used in the synthesization process. They put an explanation for that as shown in Figure 6. During the heating process, the ZnO seeds begin firstly to coalesce to form larger ZnO seeds. Those larger seeds are responsible for obtaining the larger ZnO nanowire diameters. They concluded that at low-temperature range (60–80°C), both lateral and axial growth are promoted, while at the high-temperature range ($T > 80°C$), the axial growth is mainly supported. That means the temperature plays an important role in the growth kinetic along the c-axis [0001], more than the other directions. Figure 7 represents the evolution of the average length with the growth temperature for a fixed growth time (2 hours).

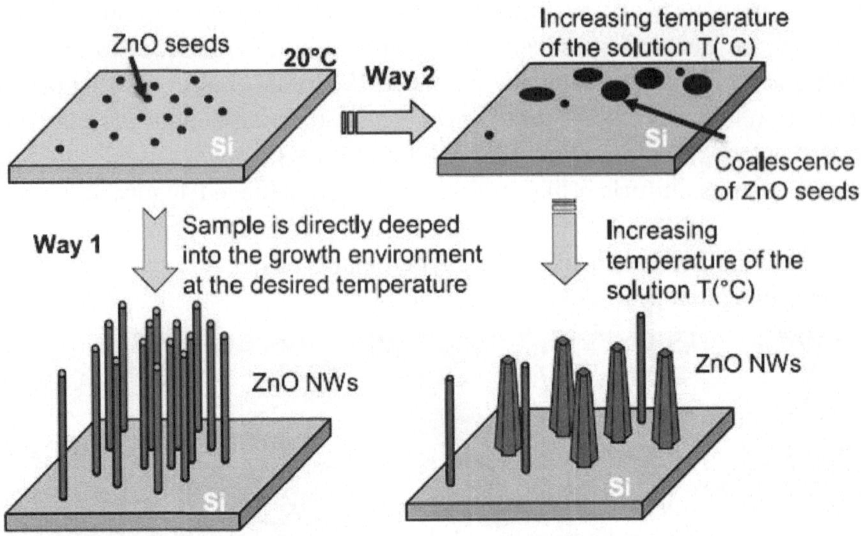

Figure 6. Schematic sketch of two possible routes for ZnO nanowire array synthesis (image from [38]).

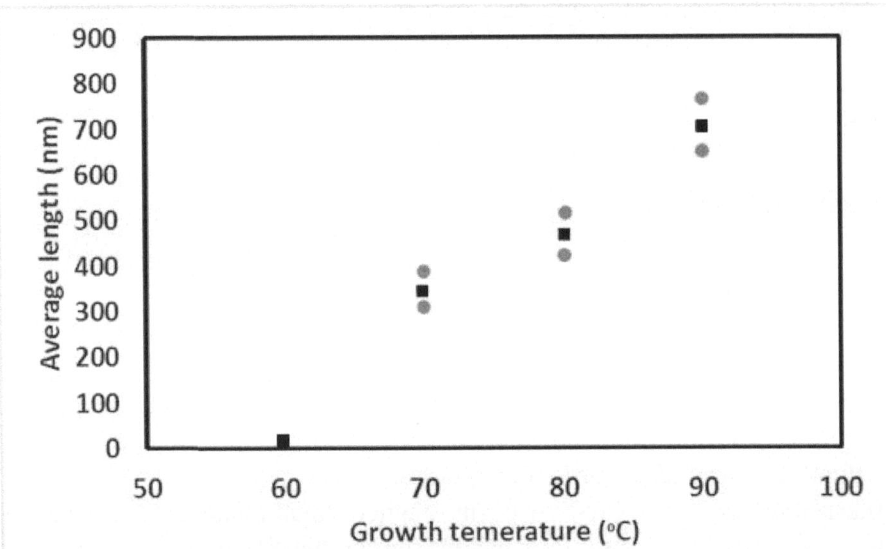

Figure 7. Average length of ZnO nanowires as a function of the growth temperature (image from [38]).

Unless Sugunan et al. did not significant any difference in the nanowires morphology for different hydrothermal growth bath temperatures carried out with low precursors concentration [31]. We found an important role of the synthesization temperature on ZnO nanowires nucleation and growth, without changing their hexagonal shape [37]. Small aspect ratio, low densities, and nonalignment hexagonal-shaped ZnO nanorods were achieved, when ZnO nanorods are synthesized at low precursor equimolar solution (5 mM) and low temperature (65°C) (Figure 8a). Elevating growth temperature to 85°C increases nanorod density but affects their shape to become prism-like and branched with growth advancement as shown in Figure 8b. As we can see from Eq. (1), seven moles of reactants produce 10 moles of products; increasing the reaction temperature leads to increase in HMTA hydrolysis and push the equilibrium forward associated with increased system entropy [39, 40]. It can be recognized that increasing temperature provides more OH− ions with mobile energy that enhance ZnO nucleation to minimize their free energy. During the growth, process nanorod shape changed from hexagonal to prism-like shape; it was probably due to increasing the electrostatic reaction between the solution ions and ZnO polar surfaces with temperature and increasing erosion reaction on the top of the nanorods.

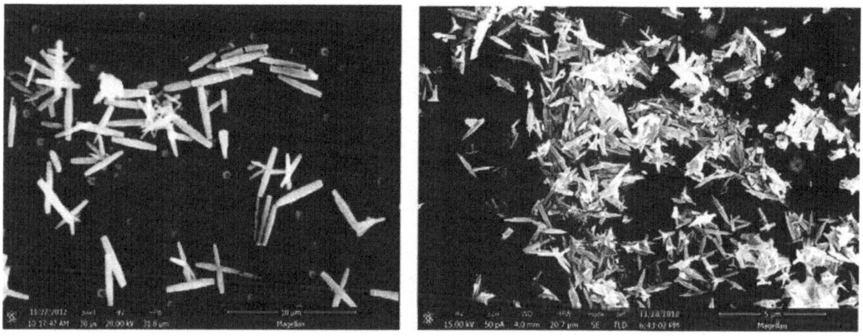

a b

Figure 8. SEM images for ZnO synthesized in (a) 65°C and (b) 85°C (images from [37]).

5.3. The seedless substrate and ZnO seeding layer

It was found that via hydrothermal synthesis method, any substrate can be used for the growth of vertical ZnO nanowires by using ZnO seeding layer. In this way, ZnO nanowires can grow on flat surface regardless of the substrate (polymer, glass, semiconductor, metal, polydimethylsiloxane (PDMS), polystyrene (PS), polyethylene terephthalate (PET), polyethylene fibers, microfibers, polyurethane, polyimide, and paper, only controlling the growth conditions). Typical pre-seeding methods include deposition of ZnO seed on the substrate by spin coating of ZnO nanoparticles, sputter deposition, and physical vapor deposition. In order to improve ZnO particle adhesion to the substrate and nanowire vertical growth, ZnO seeds from zinc acetate must be annealed at a certain temperature. The texture, thickness, and crystal size of ZnO seed layers also affect the quality of ZnO nanowire growth. Greene et al. [41] studied the minimum annealing temperature required to form textured seeds from zinc acetate on a silicon substrate. The results suggest that temperatures between 150 and 200°C are needed for seed alignment, whereas higher temperatures promote seed crystallinity and growth. While Baruah and Dutta [22] have reported that a very uniform thin layer of ZnO nanoparticles could be observed when ZnO seeds are annealed at 350°C. ZnO crystallized into nanoparticles as well as nanorod-like structures when the annealing temperature was further increased to 450°C. Ghayour et al. [42] investigated the effect of seed layer thickness on alignment and morphology of ZnO nanorods. The results showed that the diameter increased, the density decreased, and the length of the nanorods slightly decreased when the thickness of the seed layer increased (Table 1). Wu et al. [43] studied the effects of seed layer characteristics on the synthesis of ZnO nanowires. The density of nanowires decreased when the thickness increased, and the diameter of the nanowires was found to increase with the seed layer grain size. Ji et al. [44] found that the average diameter of nanowires increased and the density decreased when the seed layer thickness changed from 20 to 1000 nm. Baruah and Dutta [42] reported that the nanorods grown on seeds crystallized from a zinc acetate solution have a higher aspect ratio (of the order of 3) than those grown using nanoparticle-seeded substrates.

Table 1. The diameter, density, and length of ZnO nanorods corresponding to the thickness of the seed layer (from [42]).

Thickness of the seed layer (nm)	Diameter of ZnO nanorods (nm)	Density of ZnO nanorods (μm^{-2})	Length of ZnO nanorods (nm)
20	30	213	1052
40	36	209	1007
160	51	184	998
320	72	169	967

In seedless substrate thin layer of Au over a thin layer of Ti in order to overcome mismatching between substrate used and the ZnO particles then are annealed at 350°C under argon atmosphere. The formation of ZnO nuclei on Au surface is a complicated task to accomplish. For this, different techniques are employed to introduce the negative charges over the Au surface to attract the positively charged zinc (II) complexes, formed during Eqs. (3) and (6). It was reported that the surface roughness of the substrate has shown a major effect on the orientation and quality of the NW produced. The smooth Au (111) surface is ideal for the growth of highly aligned ZnO NWs [36].

It was found that Au provides the proper sites for nanowire ZnO nucleation on its grain boundary, Figure 9 [38].

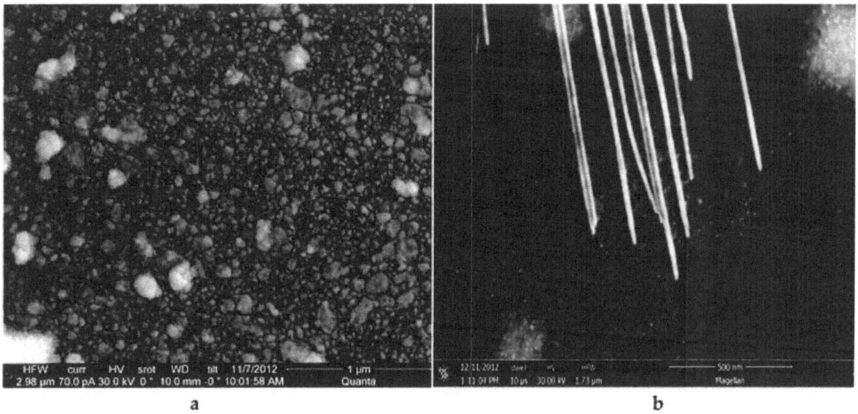

Figure 9. SEM images for seedless substrate: (a) after annealing and (b) ZnO nanowire growth on Au grain boundary. From [37].

5.4. Growth solution pH value

Recently, it was found that the introduction of ammonium hydroxide into the growth solution as an additive could efficiently affect ZnO nanowire morphology and their aspect ratio on pre-seeding and seedless substrate. Baruah and Dutta [45] synthesized ZnO nanorods on pre-seeded glass substrates using equimolar concentration of zinc nitrate and HTMA as the precursors. Flower petal-like ZnO nanostructures were obtained when the growth process is initiated in basic condition (pH 8–12). Akhavan et al. [46] studied the effect of pH change in basic range on the growth of ZnO nanorods by using zinc nitrate and NaOH as precursors. They reported that the diameters of the nanowires are increased with pH until they formed a ZnO film when the pH value reached 11.44 with a fast growth rate of ZnO nanorods on the seeded layer.

On seedless Au substrate, Boubenia et al. [36, 47] controlled both ZnO NW density and their electrical properties via developed low-cost and scalable bottom-up hydrothermal growth process using NH4OH addition. They found that small increase in pH value leads to increase the NW densities, Figure 10. They put an explanation that depends on the formation of Zn (II) complexes in the solution, which will largely decrease the Zn solubility and lead to a supersaturation large enough to initiate a large number of ZnO nuclei both in the solution and onto the Au-coated substrate. They reported that NH4OH addition has no effect on preferred growth direction of the resultant, a perfect hexagonal prism-shaped NW, and their arrays exhibit a strong preferential c-axis orientation vertically to the growth substrates.

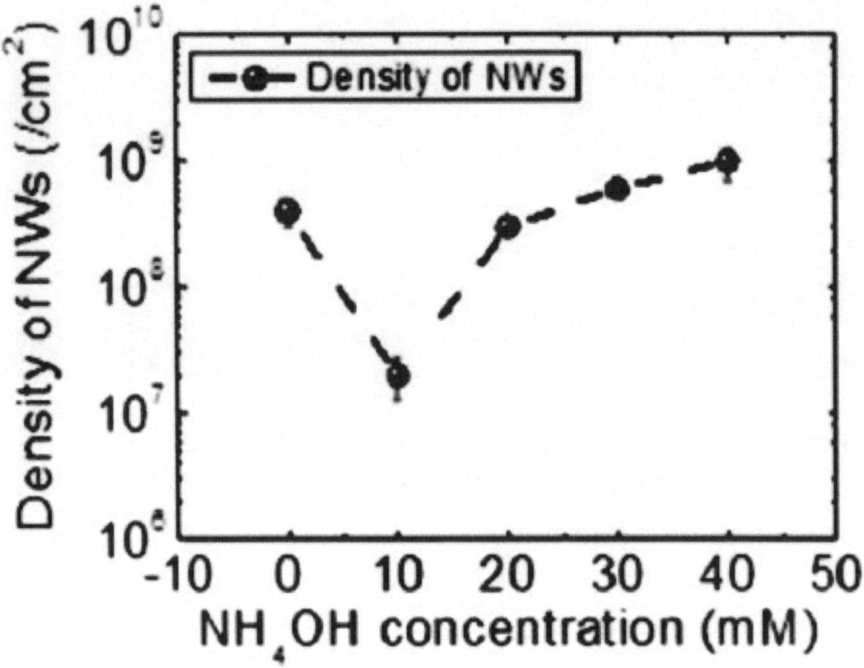

Figure 10. The variation of ZnO NW density with the change in NH4OH concentration (from [36]).

In order to adjust the pH value of the growth solution on a seeded substrate for 2 hours at 90°C, Cesar et al. [38] added HCl or NaOH to the growth solution. They observed that the ZnO nanowires are well oriented quasi-perpendicularly to the substrate and have a quite homogeneous diameter distribution for all the pH values used. But they found that pH affected ZnO nanowire morphologies. Different morphologies could be obtained: hexagonal nanorods for pH = 7, double cone-shaped tip nanorods for pH = 8, nanoprisms for pH = 10, and nanoflowers for pH = 12. For the pH value in the acid range (pH < 7.0), an attack on the seed layer leads to an inhomogeneous surface.

Igori Wallace [30] confirmed the ZnO nanowire shape changing with pH raising from 8.1 to 9 and 10 from hexagonal shape to tapered flower-like shape, respectively. They referred this change to the competition between growth and erosion. As is well known, the hexagonal wurtzite ZnO crystal is a typical polar crystal with a dipole moment in the direction of c-axis; the 0001 crystal plane represents the polarity and is metastable, but the side planes are nonpolar and relatively stable. The metastable polar top planes are able to attract hydroxyl ion OH− from the solution which contributes to the nanowire growth. Upon the addition of aqueous ammonia, OH− was increased, and its total amount was not consumed completely during the growth. The rest of the OH− in solution took part in the erosion reaction at the same time. So the relative erosion process became intensive as the pH was increased.

In our previous work [37], we found that NH4OH addition to the equimolar growth solution could change nanostructure shape and its growth direction on a seedless substrate using the hydrothermal process at 85°C. Adding NH4OH to low-concentration nutrient solution (5 mM) changed ZnO nanostructure shape from prism-like to hexagonal shape as shown in Figure 11a and b. While adding NH4OH to (10 mM) concentration nutrient solution, as we mentioned in the effect of precursor concentration, increase aspect ratio, nanorod density, and change their shape to typical hexagonal dumbbell-like, Figure 12a and b. XRD spectrums confirm the change in ZnO dominate growth surface from (002) to (112) under the addition of NH4OH. ZnO grown on <111> oriented nano-gold layer has wurtzite structure along high energy polar surfaces and XRD spectrums (Figure 2c) also. NH4OH addition leads to reduce HMT hydrolysis and Zn(OH)4−2 formation (Eq. (5) while tended to increase reaction no. 7 which means increasing growth instead of nucleation. Branching of nanorods appears when a distortion in the nanorods takes place due to stress induced with the locally increased pH value near the interface between nanorods and solution that repulse Zn ions electrostatically in order to reduce the surface energy from this distortion.

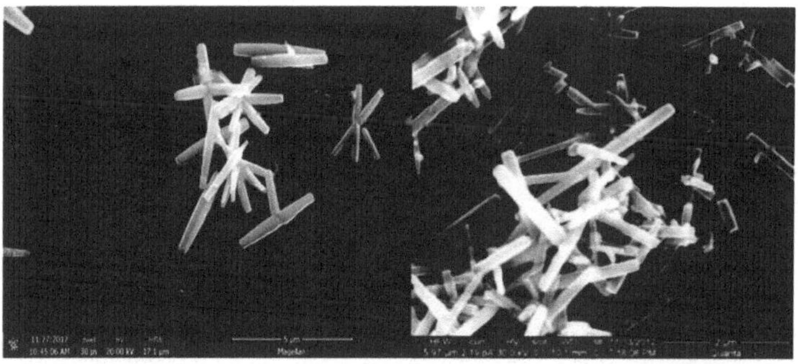

a b

Figure 11. SEM images for ZnO shape change using equimolar growth solution 5 mM (a) without addition and (b) with the NH4OH addition (image from [37]).

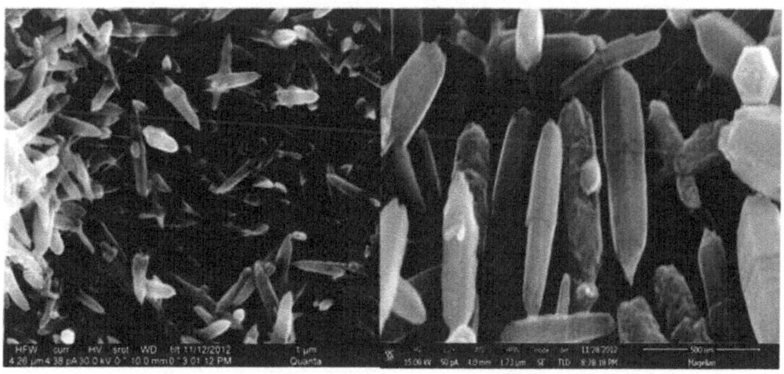

a b

Figure 12. SEM images for ZnO shape change using equimolar growth solution 10 mM (a) without addition and (b) with the NH4OH addition (image from [37]).

Increasing nutrient to HMTA ratio to 2:1 (25 mM nutrient, 12.5 mM, and low amount of 0.35 mM NH4OH) creates clear growth solution. Increasing Zn ion in the growth solution leads to fasten its precipitation as ZnO clusters on Au grain boundary of seedless substrate, Figure 12a and c. These clusters form very small-diameter nanowires grown fast as a forest-like structure. The fast depletion of Zn ions from the solution leads to increase OH− and then pH value of the solution and at ZnO/solution interface which prompts the branching process. Increasing the amount of NH4OH in the above solution creates a milky-like solution producing high-density vertically aligned ZnO nanorods (Figure 12b), due to the effect of NH4OH in controlling HMT hydrolysis and fasting ZnO precipitation from the solution. With time Zn ion depletion increased which leads to increase in OH− and pH value creating very small-diameter nanowires with nanorods that eroded according to long-time exposure to alkaline media.

5.5. Growth duration time

Yuan et al. [23] reported that the average diameter and length of ZnO nanowires increased while the aspect ratio reduced with growth duration time till slowed down due to depletion of the precursor when ZnO nanowires are synthesized using equimolar (50 mM) zinc nitrate and HMTA at 93°C (Figure 13).

Figure 13. SEM images for ZnO nanorod morphology fabricated at 25 mM nutrient, 12.5 mM HMTA, and 0.35 mM NH4OH: (a) clear, (b) milky, and (c) low magnification of clear (images from [37]).

Baruah and Dutta [22] carried out ZnO nanowire experiments on zinc acetate-seeded substrates for different growth durations from 5 to 15 hours also confirmed increasing the length and diameter of the ZnO nanowires with time. The nanowire growth slowed down after a certain period; to keep up the growth rate, they supply fresh solution to the growth solution (Table 2).

Table 2. Average diameter of ZnO nanowire changes with growth time (data from [23]).

Growth time (h)	Average diameter (nm)
0.5	35
1	37
1.5	55
2	90
2.5	100
3	100

Cesar et al. [38] investigated the effect of the growth time on the ZnO nanowire morphology synthesized at 90°C with pH of 7.0 for a growth time from 30 min to 6 hours. There are two growth stages for the synthesis of ZnO nanowires: the first stage of about 13 nm/min growth rate until 1 hour and the second one with a slower growth rate of about 2.5 nm/min, the growth time situated between 1 and 6 hours (Figure 14).

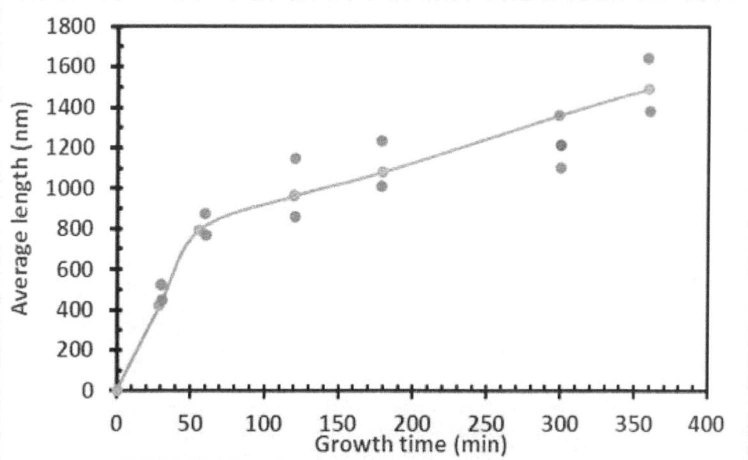

Figure 14. Average length of ZnO nanowires as a function of the growth time (from [38]).

ZnO nanostructure morphology would not be affected by increasing duration growth time when it was synthesized at low equimolar solution and growth temperature of 65°C. Activation energy is required to enhance the HMT hydrolysis and ZnO precipitation from the solution; it needs higher growth temperature.

Extended duration growth time with rising growth solution temperature to 85°C leads the prism-like ZnO nanorods to branch. Advance growth time with NH4OH addition to 5 mM nutrient solution at 85°C changed the nonaligned hexagonal nanorods to a branched nanorod as shown in Figure 15a. Increasing the growth solution concentration to 10 mM equimolar precursor's ratio, at 85°C solution temperature, leads the prism-like-shaped nanorods to branch early; lateral growth direction increased until getting continuous film after 24 hours as shown in Figure 15b. Advance growth with NH4OH addition changes the dominant growth surface to (112) from the first growth hours as shown in XRD spectrum; this replicated from nanorod shape change to a typical hexagonal dumbbell-like one.

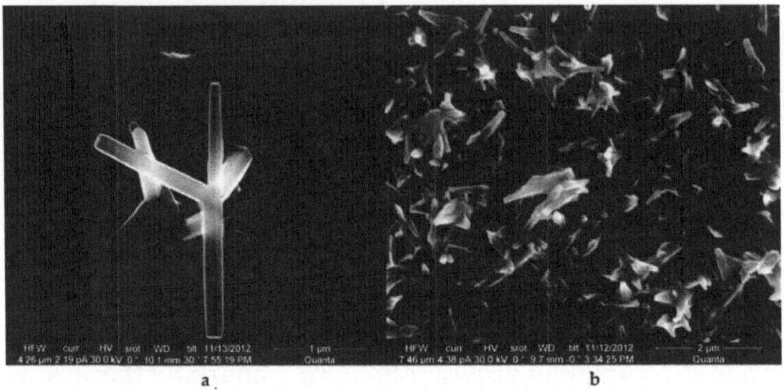

Figure 15. SEM images for ZnO nanostructure branched after long-time growth (a) with NH4OH addition to 5 mM nutrient solution at 85°C and (b) without NH4OH addition to 10 mM nutrient solution at 85°C (image from [37]).

Increasing growth time to samples in 10 mM nutrient solution at 85°C NH4OH addition leads to increase density and particle size as shown in Figure 16.

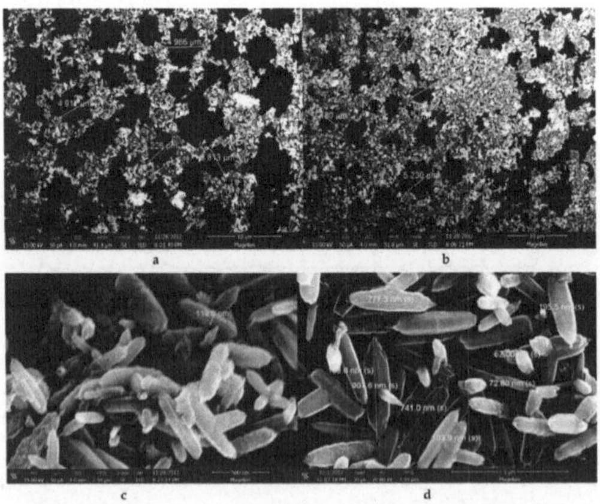

Figure 16. SEM images for ZnO nanostructure density and size for different growth NH4OH addition to 10 mM nutrient solution at 85°C: (a) 10 hours, (b) 24 hours, (c) 10 hours, and (d) 24 hours (image from [37]).

With growth duration time, NH4OH addition could prompt heterogeneous growth on the substrate, while the homogeneous nucleation in the bulk solution is suppressed so the growth could last for a long time without replenishing the solution [39]. Under this environment, an expectation of a local rise in pH value with growth time may assist initiation of more secondary nucleate branches on nanorod side surface [48]; this may explain the change in ZnO nanorod dominant growth surface and branching with NH4OH addition to the growth solution.

Long growth time for samples fabricated at 25 mM nutrient, 12.5 mM HMTA, and 0.35 mM NH4OH milky color leads to destruction of the ZnO nanowire structure due to long exposure to alkaline media, and increasing pH value in regions close to nanowires would erode their structure as shown in Figure 17.

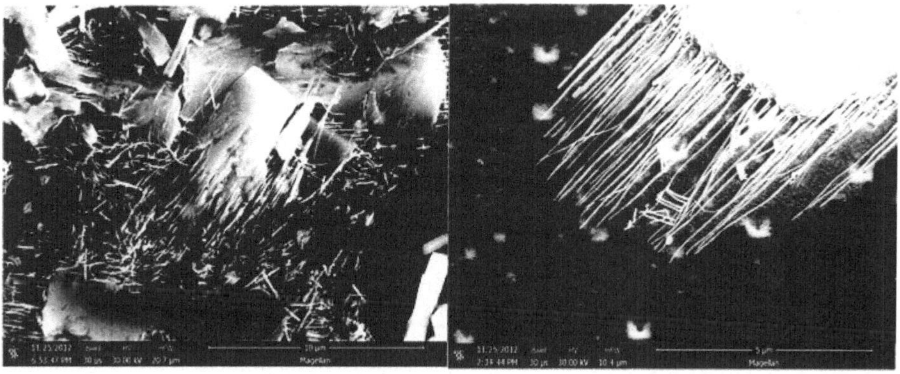

Figure 17. SEM images for ZnO nanorod morphology fabricated at 25 mM nutrient, 12.5 mM HMTA, and 0.35 mM NH4OH milky color after long growth time (images from [37]).

6. Conclusion

This book chapter provides an overview of the ZnO nanowire synthesis by hydrothermal process, the simple and efficient method that has received increased attention. A mixture of zinc nitrate and hexamine as precursor is the most popular precursor used. In this chapter we tried to focus on the parameters affecting its morphology. We reported the effect of precursor type and concentration, pH of the growth solution, bath temperature, and duration time. It is a review from published works with our experience in ZnO synthesis by hydrothermal method.

In brief conclusion, increasing the precursor concentration and growth temperature could increase nanowire density and change their morphology. Precursor concentration has the ability to increase their sizes too. Growth starting temperature influenced the nanowire alignments and also affected the density and nanowire morphology. ZnO nanowires could be synthesized via hydrothermal synthesis method on any substrate, seeded or seedless. The thickness of these seed layers affects their diameter and density. Increasing duration growth time also could increase the diameter and length of the nanowires. NH4OH addition to the growth solution changes its pH value that could increase density and change nanostructure shape and its growth direction on a seeded and seedless substrate.

References

1. Wang ZL. Venturing in ZnO nanostructures: From discovery to scientific understanding and to technology applications. Chinese Science Bulletin. 2009;54(22):4021-4034
2. Lu F, Cai W, Zhang Y. ZnO hierarchical micro/nanoarchitectures: Solvothermal synthesis and structurally enhanced photocatalytic performance. Advanced Functional Materials. 2008;18(7):1047-1056
3. Klingshirn CF. ZnO: Material, physics and applications. Chemphyschem. 2007;8(6):782-803
4. Chu S, Wang G, Zhou W, et al. Electrically pumped waveguide lasing from ZnO nanowires. Nature Nanotechnology. 2011;6(8):506-510
5. Na JH, Kitamura M, Arita M, Arakawa Y. Hybrid pn junction light-emitting diodes based on sputtered ZnO and organic semiconductors. Applied Physics Letters. 2006;95(25):53303
6. Sudhagar P, Kumar RS, Jung JH, et al. Facile synthesis of highly branched jacks-like ZnO nanorods and their applications in dye-sensitized solar cells. Materials Research Bulletin. 2011;46(9):1473-1479
7. Wang ZL, Yang R, Zhou J, et al. Lateral nanowire/nanobelt based nanogenerators, piezotronics and piezophototronics. Materials Science and Engineering R. 2010;70(3-6):320-329
8. Xu S, Lao C, Weintraub B, Wang ZL. Density controlled growth of aligned ZnO nanowire arrays by seedless chemical approach on smooth surfaces. Journal of Materials Research. 2008;23(8):2072-2077
9. Lu CY, Chang SJ, Chang SP, et al. Ultraviolet photodetectors with ZnO nanowires prepared on ZnO: Ga/glass templates. Applied Physics Letters. 2006;89(15):153101
10. Cho S, Kim S, Jang JW, et al. Large-scale fabrication of sub-20-nm-diameter ZnO nanorod arrays at room temperature and their photocatalytic activity. Journal of Physical Chemistry C. 2009;113(24):10452-10458
11. Sapkota A, Anceno AJ, Baruah S, Shipin OV, Dutta J. Zinc oxide nanorod mediated visible light photoinactivation of model microbes in water. Nanotechnology. 2011;22(21):215703
12. Liu JP, Huang XT, Li YY, Ji XX, Li ZK, He X, et al. Vertically aligned 1D ZnO nanostructures on bulk alloy substrates: Direct solution synthesis, photoluminescence, and field emission. Journal of Physical Chemistry C. 2007;111:4990-4997
13. Liu B, Zeng HC. Hydrothermal synthesis of ZnO nanorods in the diameter regime of 50 nm. Journal of the American Chemical Society. 2003;125(15):4430-4431

14. An G, Sun Z, Zhang Y, et al. CO2-mediated synthesis of ZnO nanorods and their application in sensing ethanol vapor. Journal of Nanoscience and Nanotechnology. 2011;11:1252-1258

15. Na JH, Kitamura M, Arita M, Arakawa Y. Hybrid pn unction light-emitting diodes based on sputtered ZnO and organic semiconductors. Applied Physics Letters. 2009;95(25):253303

16. Xu J, Han J, Zhang Y, Sun Y, Xie B. Studies on alcohol sensing mechanism of ZnO based gas sensors. Sensors and Actuators, B. 2008;132(1):334-339

17. Lu CY, Chang SJ, Chang SP, et al. Ultraviolet photodetectors with ZnO nanowire sprepared on ZnO: Ga/glass templates. Applied Physics Letters. 2006;89(15):153101

18. Sapkota A, Anceno AJ, Baruah S, Shipin OV, Dutta J. Zinc oxide nanorod mediated visible light photo inactivation of model microbes in water. Nanotechnology. 2011;22(21):215703

19. Wang X, Wang W, Liu P, Wang P, Zhang L. Photocatalytic degradation of E. coli membrane cell in the presence of ZnO nanowires. Journal Wuhan University of Technology, Materials Science Edition. 2011;26(2):222-225

20. Srinivasan SS, Wade J, Stefanakos EK, Goswami Y. Synergistic effects of sulfation and co-doping on the visible light photocatalysis of TiO2. Journal of Alloys and Compounds. 2006;424(1-2):322-326

21. Kim AR, Lee J-Y, Jang BR, Lee JY, Kim HS, Jang NW. Effect of Zn2+ source chemistry and materials research concentration on hydrothermally grown ZnO nanorods. Journal of Nanoscience and Nanotechnology. 2011;11:6395-6399

22. Baruah S, Dutta J. Effect of seeded substrates on hydrothermally grown ZnO nanorods. Journal of Sol-Gel Science and Technology. 2009;50(3):456-464

23. Yuan Z, Yu J, Wang N, Jiang Y. Well-aligned ZnO nanorod arrays from diameter-controlled growth and their application in inverted polymer solar cell. Journal of Materials Science: Materials in Electronics. Nov 2011;22(11):1730-1735

24. Al-Harbi T. Hydrothermal synthesis and optical properties of Ni doped ZnO hexagonal nanodiscs. Journal of Alloys and Compounds. 2011;509(2):387-390

25. Wang J, Gao L. Wet chemical synthesis of ultralong and straight single-crystalline ZnO nanowires and their excellent UV emission properties. Journal of Materials Chemistry. 2003;13(10):2551-2554

26. Gong L, Wu X, Chen H, Qu F, An M. Synthesis of vertically aligned dense ZnO nanowires. Journal of Nanomaterials. 2011;2011:5. Article ID: 428172

27. Hu H, Huang X, Deng C, Chen X, Qian Y. Hydrothermal synthesis of ZnO nanowires and nanobelts on a large scale. Materials Chemistry and Physics. 2007;106(1):58-62

28. Govender K, Boyle DS, Kenway PB, O'Brien P. Understanding the factors that govern the deposition and morphology of thin films of ZnO from aqueous solution. Journal of Materials Chemistry. 2004;14(16):2575-2591

29. Tian JH, Hu J, Li SS, et al. Improved seedless hydrothermal synthesis of dense and ultralong ZnO nanowires. Nanotechnology. 2011;22(24):245601
30. Wallace I, Eshu OV, Chukwunonso OB, Okoro UC. Synthesis and characterization of zinc oxide (ZnO) nanowire. Journal of Nanomedicine and Nanotechnology. 2015;6(5):1000321. DOI: 10.4172/2157-7439.1000321
31. Sugunan A, Warad HC, Boman M, Dutta J. Zinc oxide nanowires in chemical bath on seeded substrates: Role of hexamine. Journal of Sol-Gel Science and Technology. 2006;39(1):49-56
32. Weintraub B, Zhou Z, Li Y, Deng Y. Solution synthesis of one-dimensional ZnO nanomaterials and their applications. Nanoscale. 2010;2(9):1573-1587
33. Wang SF, Tseng TY, Wang YR, Wang CY, Lu HC, Shih WL. Effects of preparation conditions on the growth of ZnO nanorod arrays using aqueous solution method. International Journal of Applied Ceramic Technology. 2008;5(5):419-429
34. Kim AR, Lee J-Y, Jang BR, Lee JY, Kim HS, Jang NW. Effect of Zn2+ source concentration on hydrothermally grown ZnO nanorods. Journal of Nanoscience and Nanotechnology. 2011;11:6395-6399
35. Gokarna A, Parize R, Kadiri H, Nomenyo K, Gilles Patriarche A, Miskac P, et al. Highly crystalline urchin-like structures made of ultra-thin zinc oxide nanowires. RSC Advances. 2014;4:47234-47239
36. Boubenia S, Dahiya AS, Poulin-Vittrant G, Morini F, Nadaud K, Alquier D. A facile hydrothermal approach for the density tunable growth of ZnO nanowires and their electrical characterizations. Scientific Reports. 2017;7:15187. DOI: 10.1038/s41598-017-15447-w
37. Al-lami S, Jaber H. Controlling ZnO nanostructure morphology on seedless substrate by tuning process parameters and additives. Chemistry and Materials Research. 2014;6(4):101. www.iiste.org ISSN 2224-3224 (Print) ISSN 2225-0956 (Online)
38. Chevalier-César C, Capochichi-Gnambodoe M, Leprince-Wang Y. Growth mechanism studies of ZnO nanowire arrays via hydrothermal method. Applied Physics A. 2013. DOI: 10.1007/s00339-013-7908-8
39. Xu S, Wang ZL. One-dimensional ZnO nanostructures: Solution growth and functional properties. Nano Research. 2011;4(11):1013-1098
40. Govender K, Boyle DS, Kenway PB, O'Brien P. Understanding the factors that govern the deposition and morphology of thin films of ZnO from aqueous solution. Journal of Materials Chemistry. 2014;14:2575-2591
41. Greene LE, Law M, Tan DH, et al. General route to vertical ZnO nanowire arrays using textured ZnO seeds. Nano Letters. 2005;5(7):1231-1236
42. Ghayour H, Rezaie HR, Mirdamadi S, Nourbakhsh AA. The effect of seed layer thickness on alignment and morphology of ZnO nanorods. Vacuum. 2011;86(1):101-105

43. Wu WY, Yeh CC, Ting JM. Effects of seed layer characteristics on the synthesis of ZnO nanowires. Journal of the American Ceramic Society. 2009;92(11):2718-2723

44. Ji LW, Peng SM, Wu JS, Shih WS, Wu CZ, Tang IT. Effect of seed layer on the growth of well-aligned ZnO nanowires. Journal of Physics and Chemistry of Solids. 2009;70(10):1359-1362

45. Baruah S, Dutta J. pH-dependent growth of zinc oxide nanorods. Journal of Crystal Growth. 2009;311(8):2549-2554

46. Akhavan O, Mehrabian M, Mirabbaszadeh K, Azimirad R. Hydrothermal synthesis of ZnO nanorod arrays for photocatalytic inactivation of bacteria. Journal of Physics D. 2009;42(22):225305

47. Dahiya AS, Boubenia S, Franzo G, Poulin-Vittrant G, Mirabella S, Alquiera D. Photoluminescence Study of the Influence of Additive Ammonium Hydroxide in Hydrothermally Grown ZnO Nanowires. Posted: 16 March 2018. Preprints (www.preprints.org). DOI: 10.20944/preprints201803.0138.v1

48. Zhang TR, Dong WJ, Keeter-Brewer M, Konar S, Njabon RN, Tian ZR. Site-specific nucleation and growth kinetics in hierarchical nanosyntheses of branched ZnO crystallites. Journal of the American Chemical Society. 2006;128:10960-10968

5

Nanowires for Room-Temperature Mid-Infrared Emission

1. Introduction

Developments that took place in the past few decades in the semiconductor industry have allowed the realisation of III–V one-dimensional (1D) structures [1], such as nanowires (NWs), that have attracted increasing attention as promising materials for the fabrication of mid-infrared nanoscale devices [2]. III–V semiconductors NWs have many interesting physical and optical properties due to their narrow band gap [3], small electron effective mass [4], very high electron mobility [5], along with a great potential for realising nanoscale devices [1, 6, 7].

In recent years, silicon photonics attracted a significant research effort because of the potential benefits of integrating optoelectronics functions within silicon (Si) CMOS electronic devices [8]. The growth of InAs nanowires on low-cost Si substrates paves the way towards low-cost infrared detection technologies [7]. The unique geometry of NWs structures offers new silicon photonics architectures for sensing to be operating in the mid-infrared spectral range [9], hence improving the control over the size [10], higher flexibility in sample processes [11, 12] and more freedom for band-gap engineering [13]. The field of InAs-based NWs growth in particular has attracted many researchers interest and have been extensively investigated in order to fabricate nanoscale devices including field-effect transistors [14], solar cells [15], sensor applications [16], lasers [12] and photodetectors [17].

Semiconductor NWs in nanotechnology can be synthesised through two main approaches, one is called top-down and the other bottom-up approaches [18]. The idea behind the top-down approach is to etch out and remove the crystal planes of the material from larger pieces which is usually already present on the substrate to form the nanowires. This approach mostly dominates in industry for large-scale fabrication. Few researches demonstrated successful production of NWs using this approach, for example, using InP and InGaAsP/InP materials [11, 19]. To the best of our knowledge, until today, there has not been any report of InAsSb fabricated using top-down approach; this approach shows to some extent the ability to be producing NWs for some III–V materials. However, this method suffers from drawbacks such as wire surface contamination or damages after etching treatment which hinders achieving an optimum optical performance. In the bottom-up approach, the nanostructures are built up on the substrate by adding atoms layer by layer in an ordered manner, offering the growth of a very high uniformity to the crystal structures, with a higher relative controllability in the NWs growth rate. There are two methods within the bottom-up fabrication of nanowire growth techniques: the template directed and the free-standing methods. Most of the reported NWs have been fabricated using the second approach by random or site-controlled growth.

The NWs growth mechanisms are usually governed by the famous vapour-liquid-solid (VLS) [20] or vapour solid (VS) mechanisms [21]. In the VLS approach, metal droplets are deposited on the growth substrate either through self-induced (e.g. In for InAs) or foreign metal catalyst (such as Au droplets) followed by subsequent nanowire synthesis. In the VS growth mechanism, the material starts in a vapour form inside the growth chamber and then the layers are deposited on top of the substrate epitaxially layer by layer in the solid phase, and such processes can be lithographically patterned or self-assembled. In the case of self-assembled, the NWs are grown randomly on the surface and have a variation of diameter and length. However, this kind of growth may lead to an unintentional kinking in the grown NWs [14]. Thus, controlling over position and size (diameter and length) of the NWs are required to fabricate efficient nanowire-based devices, hence allowing their applications to be realised at a large scale.

As more progress is made towards realising efficient optoelectronic devices, many methods are adapted for NWs growth, such as pulsed laser deposition [22], chemical beam epitaxy [23], metal organic chemical vapour deposition MOCVD [24] and molecular beam epitaxy (MBE) [25]. The most popular methods to grow NWs are (MOCVD) [26] and (MBE) [27]. MOCVD system and related gas phase techniques are used in the production of commercial large-scale product, due to their high surface diffusion, fast growth rates and high throughput [28]. However, in comparison to MOCVD, solid-source MBE offers several advantages in low impurity incorporation due to the ultrahigh vacuum growth environment and the highly pure elemental growth species [26], the very accurate control over the composition of the deposited monolayer and their doping, and finally the ability to grow advanced radial and axial core-shell heterostructures via sophisticated in situ growth-monitoring methods [29].

The focus in earliest studies was on optimising the growth conditions, analyse or control the crystal structure for such wires. However, the crystal phase is particularly relevant to PL studies because the band gap and hence emission wavelength are phase dependent. In general, bulk InAs grown by conventional epitaxial techniques have a zinc-blende phase, hence its band gap (EgInAs (ZB)) is well characterised at 0.415 eV. However, by contrast, a consensus has not yet been reached around a single band gap for Wurtzite InAs (EgInAs (Wz)). While low-temperature studies build initial understanding, only a few optical emission studies have been reported, due to the very poor optical efficiency of these materials, very strong non-radiative surface and difficulties of performing spectroscopy in the IR spectral region [30]. For instance, the first low-temperature PL of InAs NWs was reported by Sun et al. [30] for InAs NWs on Si having both pure WZ and ZB crystal phases, with band energy (0.41–0.425 eV) corresponding to the above band edge surface state-related recombination, with a slight shift at increasing the temperature. They also noticed a blue shift due to quantum confinement depending on the wire diameter and not on the structure changes. Furthermore, Trägårdh et al. [31] predicted a WZ band gap of 0.54 eV from extrapolation-fitted photocurrent measurements at 5 K on single InAs NWs with a centrally placed InAs1−xPx segment of the composition $0.14 < x < 0.48$. Möller et al. [32] reported temperature-dependent PL studies that enable them to estimate the wurtzite band gap to be 0.458 eV at low temperatures. Koblmuller et al. [11] have gone further to report PL temperature dependence of InAs NWs having a WZ phase structure and a peak position 0.411 eV at 15 K. Despite the strong PL emission at low temperatures, the emission was only reached 130 K before the signal was quenched [31]. The wires show a 25-meV blue shift due to confinement by reducing the wire diameters from 100 to 40 nm with respect to bulk InAs due to quantum confinement and dominant surface effect which limits PL efficiencies. The non-radiative recombination causing this quenching could originate from lattice defects, surface states or Auger recombination [31]. Recently, Rota et al. [33] estimated the energy gap for InAs WZ nanowires to be 0.477 eV, higher than the ZB band gap by 59 meV which does not depend on the nanowires size and carrier confinement. The spread of results that ranges from 0.41 to 0.54 eV may originate from polytypism, with a further complication being atmospheric absorption in the commensurate spectral range. However, while low-temperature PL measurements have supported initial studies confirming the crystal structure, emission at room temperature will be required for most practical applications. A common route to suppressing non-radiative recombination at the surfaces is the in situ growth of a wider band-gap shell which has been employed to InAs wires by Treu et al. [34] using an InAsP shell where they demonstrated 102 times enhancement of the PL emission up to room temperature [35]. Also, GaAs/AlGaAs core-shell NWs showed improved PL intensity compared to bulk GaAs NWs caused by the reduction in the surface states, which was found to be effective in enhancing the PL emission intensity, allowing it to persist up to room temperature. In addition to suppressing loss through non-radiative recombination, PL intensities and quenching temperatures can be increased by acting to raise the radiative recombination rate. In a very recent study, Jurczak et al. demonstrated a 10-fold enhancement of InAs NWs PL emission using an InP core-shell layer that passivates the surface states to reduce the rate of non-radiative recombination [35]. This research direction is attracting many researchers today in order to develop advanced optoelectronic devices and nanoscale photonic applications [13]. Progress in this direction will provide further insight into the optical emission and energy band-gap properties, hence improving the use of these materials, especially for infrared detectors and emitters. This chapter discusses the concept of developing novel InAsSb/InAs multi-quantum wells (MQWs) NWs on Si (111) substrate structures within InAs nanowires, as significant step towards viable nano- and quantum emitters in the extended IR wavelength range. We review the growth process for these structures, the crystal structural characterisation. Finally, we discuss their optical properties along with developing a band structure for the NWs of this material.

2. The growth

The growth of high-quality nanowires should be achieved by avoiding the common growth process that employs a foreign catalyst such as gold to nucleate the wires. Au is well known to introduce deep level traps in the material band gap as contaminations [36, 37], which then limits the performance of the devices functionality on Si. Therefore, using lithographically predefined SiO2 template is an additional benefit besides avoiding catalysts which enables an accurate control over positions and diameters of the grown NWs that determines where the growth occurs and allows homogeneous arrays by controlling the nucleation position without any catalysts by the selective area epitaxy technique.

In this work, the selective area molecular beam epitaxy (SA-MBE) technique has been used as a first step to grow the InAsSb/InAs MQWs NWs on Si (111) wafers and was achieved with nano-hole patterns produced by EBL [38]. The substrate is masked with a patterned dielectric layer, normally a SiO2 layer with a thickness below 100 nm. The NW growth starts when the As and In adatoms start to form critical nuclei of certain sizes on the surface inside the patterned holes.

The MQW nanowire growth was initiated by impinging As flux followed by exposure of the sample to the In flux 20 s later. After the growth of an initial pure InAs section for 1 h, the MQW active region was grown as 10 repeats of InAs/InAs1−xSbx MQWs with growth durations of 180 and 27 s, respectively. This was expected to form 10 repeats of 25 nm InAs and 8 nm InAs1−xSbx, giving a total active region thickness of 330 nm as shown in Figure 1(a).

Finally, the wires were finished with an InAs cap, grown for 10 min. All growths began with a pure InAs section and were finished with an InAs cap. For further comparison, a planar bulk InAs sample was grown as a reference under the same growth conditions. The MQW NWs were grown in arrays of 50-nm diameter holes patterned using e-beam lithography in a 330-nm pitch square array defined within a number of 200 × 200 μm areas, on each silicon substrate, and the final hole diameter was ~80 nm due to lateral etching. From the SEM images, the MQW wires were on average ~1.5 μm in length and ~100 nm in diameter, see Figure 1(b) and (c).

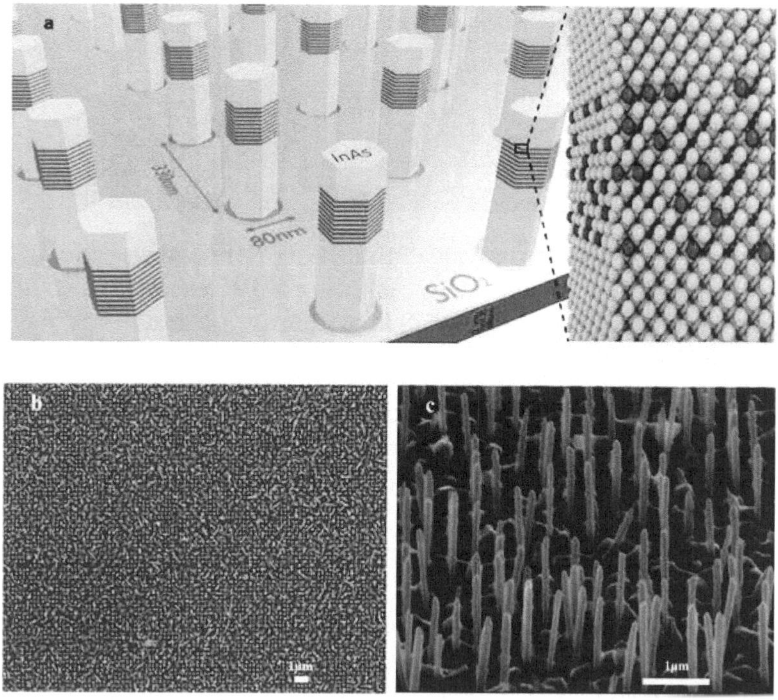

Figure 1. Nanowire structural design and analysis. (a) Schematic drawing of the InAsSb/InAs MQW within InAs NWs, grown in a 330-nm pitch square array patterned on a silicon substrate using a Si02 template with 80-nm diameter holes. Blue and yellow spheres represent In and As atoms, respectively, while red spheres indicate the random incorporation of Sb within the quantum well region. (b and c) Top view and tilted view SEM images, respectively. The nanowires are viewed at an angle of 70°. The scale bars in (b) and (c) are 1 μm. Figure obtained with permission from authors [38].

3. Structural analysis

The lattice structure of the InAs/InAsSb MQW NWs has been investigated using Scanning Transmission Electron Microscopy (STEM). A close inspection of high-resolution STEM images shows both the angled nature of the well and a continued WZ structure within the InAsSb well. Figure 2 shows a full diameter image of the wire together with a zoomed-in view, within which the lattice structure can be discerned. From the zoomed-in view, it is clear that the structure of the InAsSb well does not change to ZB, as it would do in a pure InAsSb NW. However, in the case of the QWs, the short growth durations are insufficient to allow a flat top to form and so the WZ phase is maintained. Hence, the Sb fraction in the wells is currently best estimated at 6–7%, based on comparable bulk wires.

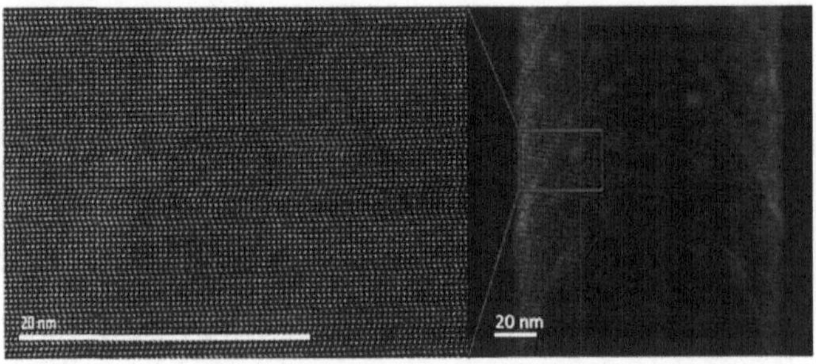

Figure 2. High-resolution TEM analysis of the QW in NW. Cross-sectional TEM images, looking across a section of wire including a QW. The lattice structure is visible, and the crystalline phase (WZ) is evidently continuous across the quantum well. Figure obtained with permission from authors [38].

The nanowires had a regular hexagonal cross section with {10 1 ¯ 0} sidewalls, faceted tips and a twinned WZ crystal structure with stacking faults, which is a characteristic of InAs NWs [39, 40]. In common with other researchers, it has been found that the addition of Sb to form bulk InAsSb NWs forces a rotation in the layer stacking, leading to a predominantly ZB structure with a flat top. Bulk InAsSb NWs grown under these conditions contained 6–7% Sb, in good agreement with the earlier work that reports saturation at this concentration [33]. However, when the InAsSb growth is limited to nanoscale quantum wells, energy-dispersive X-ray (EDXS) mapping revealed preferential incorporation of the Sb on specific crystal planes. This results in the formation of novel quantum wells having faceted, flat-topped, conical shape, with open or partially closed flat tops, as shown in Figure 3(a). It is noted that in other NW studies, the authors have considered nanoscale axial heterostructures as both quantum dots (QDs) [41, 42] and QWs [43, 44, 45]. The QD model is understandably favoured for lower-wire diameters; however, in light of the larger 100-nm diameter and the weak lateral conferment of the higher effective mass holes, the heterostructures described here are considered to be QWs.

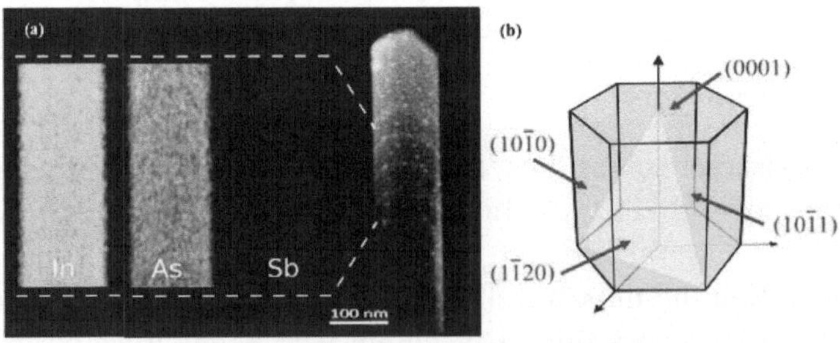

Figure 3. (a) STEM image showing the distribution of In, As and Sb obtained from 2D EDXS mapping and the resulting unusual facetted conical shape of the InAsSb MQW and (b) the unit cell of the WZ crystal structure showing the possible growth planes for the InAsSb MQW facets. Figure obtained with permission from authors [38].

4. 4-K micro-PL measurements

The 4-K PL emission from the MQW was studied using temperature-dependent micro-photoluminescence spectroscopy. Introducing the InAsSb MQW significantly changes the PL emission characteristics of the NWs in a number of ways. A comparison of the spectra measured at 4 K from the InAsSb MQW NWs, bulk alloy InAsSb NWs and InAs NW samples is shown in Figure 4.

Figure 4(a) and (b) shows the PL emission from the InAsSb MQW NWs at high and low excitations, respectively. The emission from the bulk InAsSb NWs is shown in Figure 4(c) and are deconvoluted into peaks at 0.380 eV corresponding to 6% Sb with a dominant ZB phase in agreement with previous work [46], and a shoulder on the main peak originating from ZB InAs which appears as the dominant phase in the early stage of all the NW growths [47].

The emission from the InAs NWs peaks at 0.482 eV is shown in Figure 4(d), demonstrating a dominant WZ phase [3, 33]. The PL emission of the InAsSb MQW NWs collected at high and low excitations exhibits a clear increase in peak emission energy with respect to the bulk InAsSb NWs. This is due to the strong carrier confinement within the quantum wells. In addition, the InAsSb MQW NWs also exhibit an increased emission intensity and a superior temperature-quenching behaviour compared with the bulk InAsSb NWs as expected. Most notably, the emission intensity is enhanced at all temperatures, due to the quantum confinement of electrons and holes. PL originates from type II spatially indirect recombination of electrons in the InAs layers with confined hole states in the InAsSb QWs, where the spatial separation helps in reducing non-radiative Auger recombination with a corresponding increase in radiative emission [48]. Single Gaussian fitting to the spectra reveals interference from characteristic atmospheric absorption by water vapour between 0.445 and 0.485 eV [10]. The spectra can be scaled to account for the reduced cross-sectional area of the nanowire samples, where only 7% of the surface area is covered by the NWs, assuming a 100% nucleation yield in the mask sites. Accounting for this reduced active area allows the most direct comparison of emission intensity.

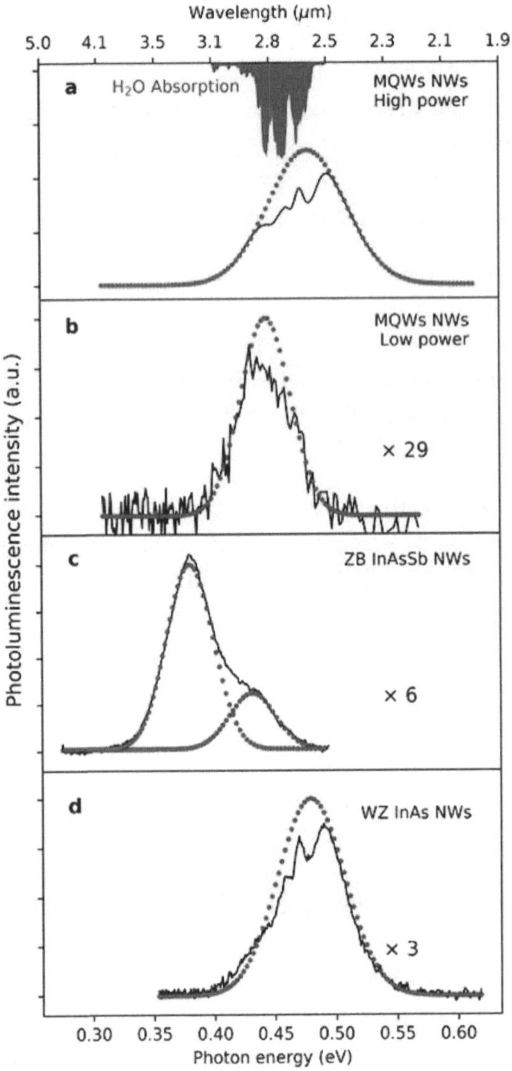

Figure 4. Micro-PL spectra obtained at 4 K. The emission spectra measured from (a) InAsSb/InAs MQW NWs at high power (1.2 × 104 W cm−2), (b) InAsSb/InAs MQW NWs at low power (80 W cm−2), (c) InAsSb NWs at (1.2 × 104 W cm−2) and (d) InAs NWs at (1.2 × 104 W cm−2). The dotted lines represent Gaussian fits to the spectra, which also reveal the atmospheric absorption from water vapour [49] in this spectral region, as shown in (a). Figure obtained with permission from authors [38].

We observed that at the low excitation (0.011 W cm−2), the MQW NW emission intensity is 4.2-fold enhanced with respect to a bulk InAs reference sample, compared to a 1.6-fold enhancement from the InAs NWs (see Figure 5). It must be noted that this comparison assumes a direct area proportionality for the optical pumping efficiency. However, it has been shown that the efficiency of optical absorption in nanowire arrays exhibits a spectral dependence arising from mode guiding, due to the geometry of the wire and the array, such that the peak field can occur either inside or outside the wires [50]. To a first order, this effect is defined by the wire diameter, and in prior work, a very similar effect has been reported from an array of InAsSb NWs in a photodetector [11], where the peak response was obtained at 1.5 μm with an FWHM of 320 nm. Consequently, there is a non-optimal coupling with the 808-nm pump laser used in the present PL studies, and hence, further enhancement of PL emission intensities is to be expected if the pump laser wavelength is correctly matched to the NW geometry.

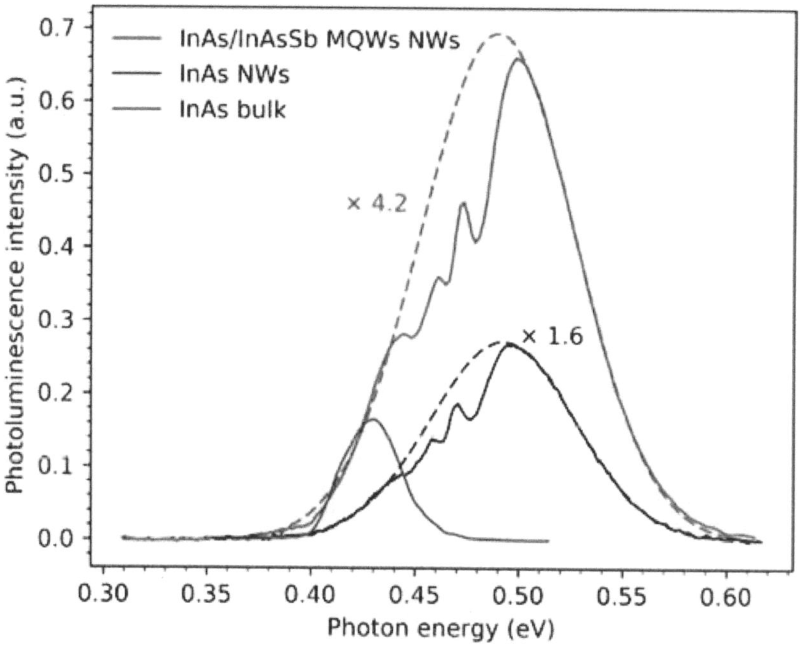

Figure 5. Comparison of emission intensities. PL spectra, under 3.2 × 104 W cm−2 excitation, showing the relative emission intensities for MQW NWs, InAs NWs and an InAs bulk sample, scaled by active area. Figure obtained with permission from authors [38].

5. PL power dependence

Under low excitation conditions, bulk ZB InAs at 4 K normally exhibit characteristic emission from bound exciton and donor-acceptor transitions around 0.417 and 0.374 eV, respectively [51]. The high excitation intensity (~ 04 W cm−2) in our micro-PL experiments results in state filling such that a single InAs peak is observed at 0.425 eV. In the present case, the InAs NW emission is further blue-shifted with respect to the bulk ZB reference sample, due to the WZ crystal structure of the NW, with a peak emission energy ranging from 0.469 eV under low excitation, to 0.485 eV under high excitation, see Figure 6. The band gap for WZ InAs is known to be higher than that of ZB InAs, and our result is consistent with earlier studies of WZ InAs NWs which reported band gaps in the range of 0.477–0.540 eV [12, 32]. Quantum confinement-induced blue shifts have also been observed as the diameter of InAs NWs is reduced [14]; however, in the case of 100-nm diameter wires, the shift is rather small ~5 meV. Hence, the emission observed from the InAs NWs grown in this work is in good agreement with earlier reports. We envisage modest size-related confinement effects in the InAsSb MQW NWs. The PL emission energy from the InAs NWs is blue-shifted by 21 meV over the range of pump powers used in our experiments, which is similar to that obtained by others [14] and is associated with band filling. The commensurate shift for the MQW NWs is much larger, at 45 meV (see Figure 6). The majority of this blue shift occurs at low pump powers before the dependence becomes similar to that for pure InAs wires at higher powers. The blue shift in the MQW nanowires arises due to band-bending effects characteristic of type II QWs and originates from Coulombic attraction between localised holes in the InAsSb quantum well attracting electrons from the adjacent InAs barrier forming triangular quantum wells.

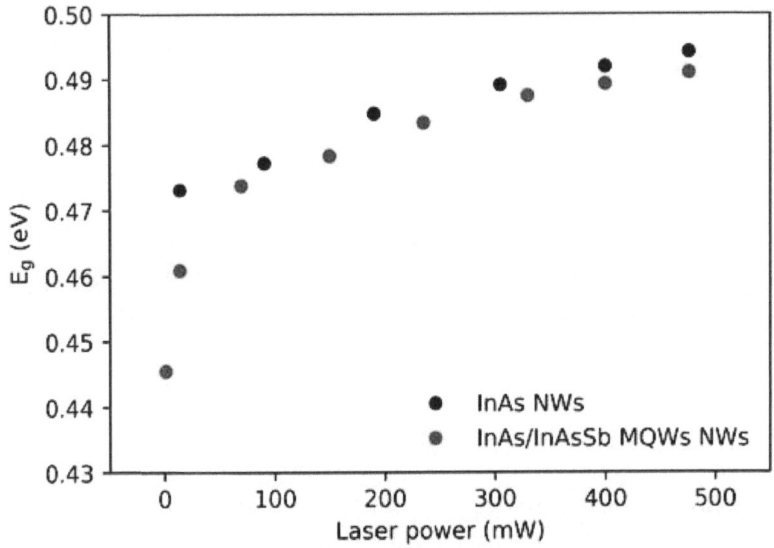

Figure 6. Power dependence of PL emission. The dependence of the peak emission energy on the power of the pump laser incident on the sample, for InAs and InAsSb/InAs MQW NWs, showing the difference in the blue shift between the pure InAs NW, with minimal quantum confinement effects and the MQW NW, with a strong quantum confinement and charging effects. Figure obtained with permission from authors [38].

An increase in excitation intensity will raise the steepness of the confining potential and consequently the electron quantisation energy E, with a typical ΔE~L1/3 behaviour. Accordingly, the flat-band transition energy of the InAsSb/InAs MQW can be extracted from the intercept of Figure 7(a)—the PL peak position versus L1/3. Consequently, the flat-band transition energy for the InAs/InAsSb MQW NWs is obtained as 0.438 eV which is in good agreement with the calculated transition energy of 0.443 eV, as shown in the schematic band diagram in Figure 7(b).

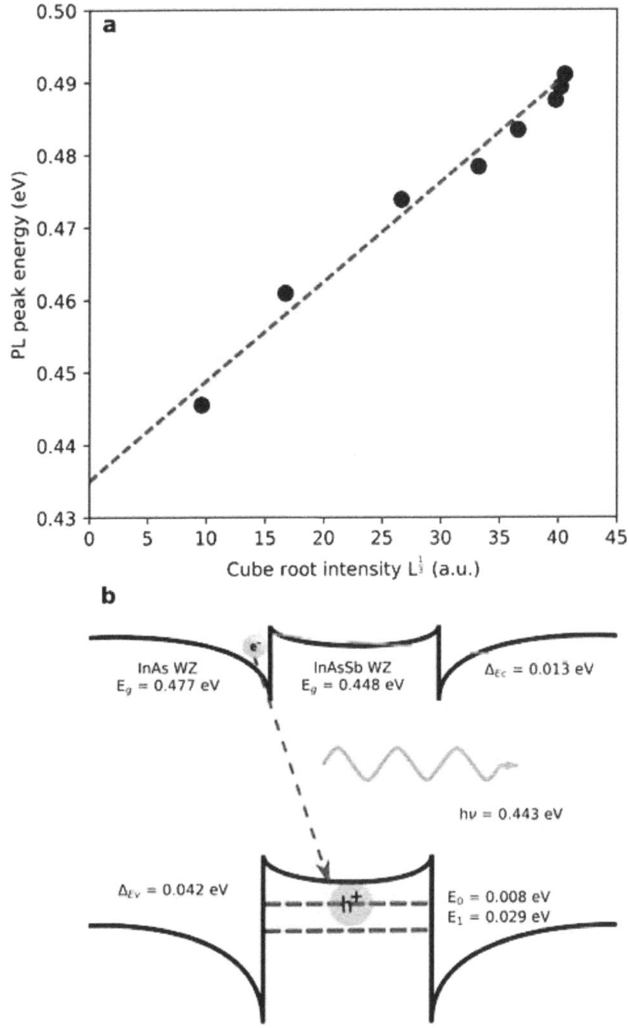

Figure 7. Determination of the flat-band transition energy at the QW. (a) PL peak energy versus cube root of integrated PL intensity, elucidating the charging of the type II QW and allowing the flat-band transition energy to be identified and (b) the calculated band diagram for the InAsSb/InAs QW and photon energy, showing the band bending and triangular well formation. Figure obtained with permission from authors [38].

6. Development of a band structure for the MQWs

Developing a band diagram for the InAs/InAsSb MQW NWs is not straightforward, due to limited data describing band gaps and alignments for the Wurtzite phases. In particular, pure InAs1−xSbx nanowires take on a predominately zinc-blende structure for x > 4%, precluding the measurement of Wurtzite band gaps at higher antimony fractions. By contrast, the InAs1−xSbx growth in this work maintains its Wurtzite structure due to the growth being on the (10 1 ¯ 1) facets of the Wurtzite InAs wire. To approximate the band diagram in this absence of reported parameters, we start with the band alignment for a comparable zinc-blende structure, calculated taking account of the Sb fraction and the strain within a (111) orientated nanowire. The band gap of Wurtzite InAs is taken to be 60 meV greater than that for InAs zinc-blende (EGwz), with the noted caveat that there is a lack of consensus in the study. The value of the band gap for the wurtzite InAs1−xSbx can be calculated using the quadratic approximation:

$$E g\, InAs\, 1 - x\, Sb\, x = x\, E\, gWZ\, InAs + 1 - x\, E\, gWZ\, InSb + x\, 1 - x\, 0.67\, E1$$

where 0.67 is the bowing parameter for zinc-blende InAsSb and EgWZ (InSb) is the band gap of wurtzite InSb taken as 0.287 eV [52]. This gives Eg (InAs0.93Sb0.07) = 0.424 eV which is 63 meV greater than the known value for the zinc-blende phase. This also agrees with results by Farrell et al. who showed that for x = 3.9%, the band gap of Wurtzite InAs1−xSbx was also ~60 meV greater than the known value for the zinc-blende phase.

The confined hole states were calculated using a six-band k.p. solver in Nextnano. The first confined heavy hole state is calculated to be 8 meV above the band edge, as shown in the schematic energy band diagram in Figure 7(b) corresponding to a flat-band recombination transition energy, Et, of 0.443 eV.

7. PL temperature dependence

PL spectra obtained at different temperatures for the InAsSb MQW NWs are shown in Figure 8. Although the wires are not capped or passivated, they exhibit strong PL emission which persists up to room temperature. This indicates that radiative recombination occurs primarily in the MQW away from the near surface regions, which in InAs NWs are known to be accumulated due to Fermi level pinning, resulting in a low efficiency for radiative emission [53]. In our case, the quantum confinement of the MQWs allows room-temperature emission to be observed without passivation. The PL spectra are inhomogeneously broadened due to length variations in the NWs which also result in thickness variations in the QWs. The atmospheric water vapour absorption is again evident near 2.7 μm (0.459 eV) in all the spectra.

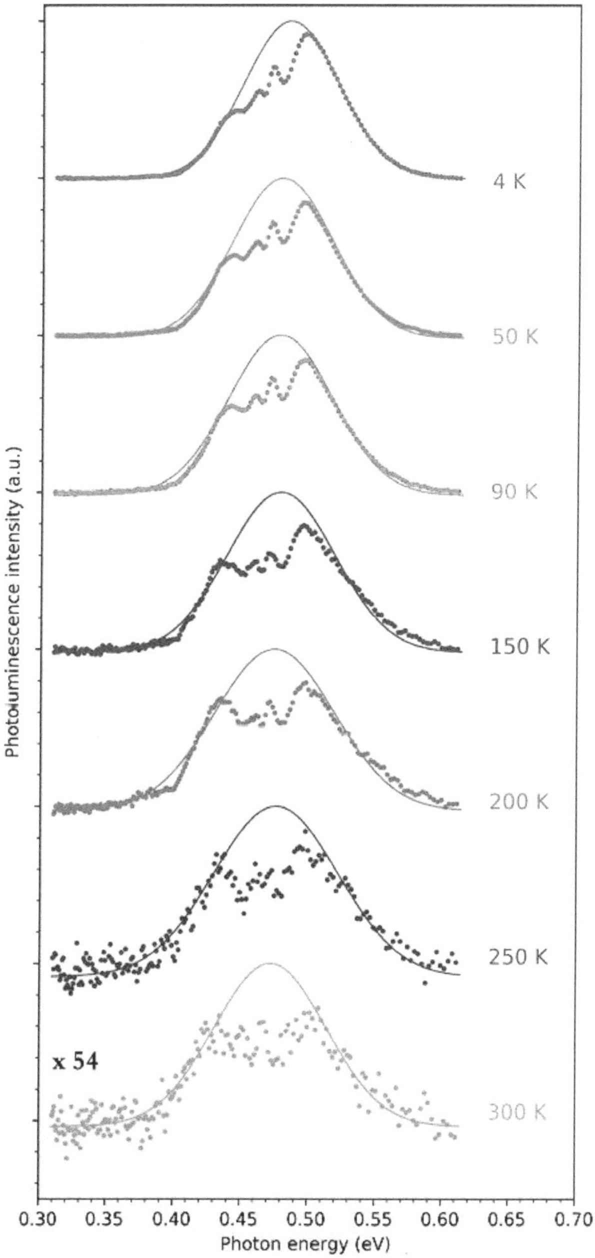

Figure 8. Temperature dependence of the PL emission spectra obtained from the InAsSb MQW NWs. Emission spectra measured over the range of 4–300 K using high excitation (2.6 × 104 W cm−2), showing the room-temperature emission required for future practical NW infrared emitters. The dotted lines indicate Gaussian fits used to extract the peak emission wavelength. Atmospheric water vapour absorption is again evident in all the spectra. Figure obtained with permission from authors [38].

Figure 9 presents a comparison of the temperature quenching of the NW samples, where the superior performance of the InAsSb MQW NWs is clearly evident due to the increase in the radiative emission rate and suppression of Auger recombination.

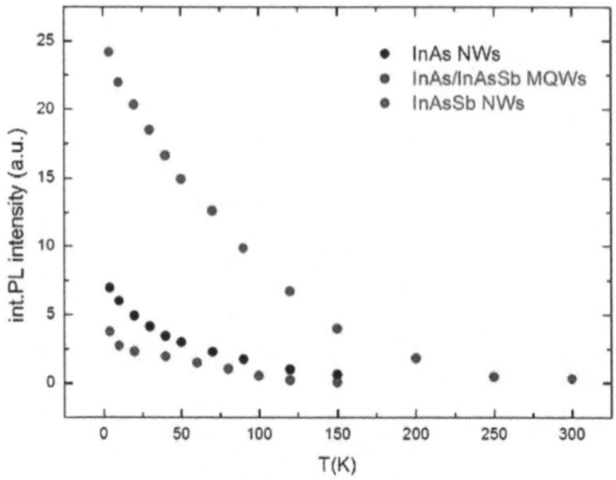

Figure 9. Temperature-dependent analysis of PL data from the InAsSb/InAs MQWs, InAsSb and InAs NW samples; the graph demonstrates quenching behaviour of the three NW sample-integrated PL intensities with increasing temperature. Figure obtained with permission from authors [38].

The temperature dependence of the peak energies for the NWs is shown in Figure 10. The dotted lines represent fitting of the results using the empirical Varshni equation. The values obtained for the fitting parameters (Eg (0), α and β) are given in Table 1 along with reference values for bulk InAs; both the InAs and the InAsSb MQW NWs have a WZ crystal structure and consequently have a weaker dependence of band gap on temperature (lower value of α) than the corresponding bulk ZB materials.

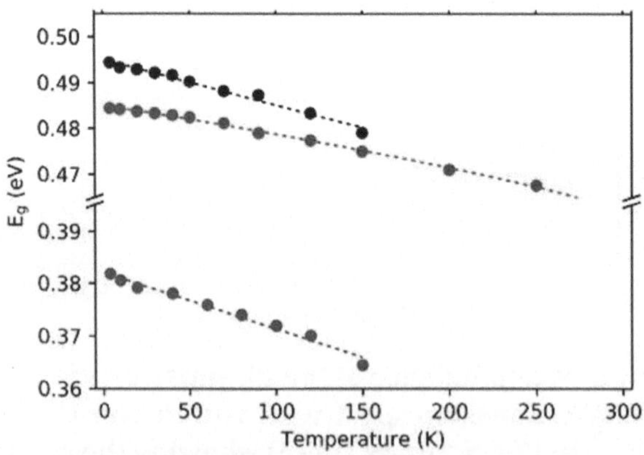

Figure 10. Temperature-dependent analysis of PL peak from the InAsSb/InAs MQWs, InAsSb and InAs NW samples. The temperature dependence of the PL peak emissions (points) was used to fit the Varshni relationships (solid lines) and extract the associated coefficients, for the three NW samples. Figure obtained with permission from authors [38].

Table 1. Comparison of Varshni parameters. The fitted Varshni parameters E0, α and β for the NW samples, compared with published parameters for bulk InAs and InSb [54].

	E0	A	B
InAs NWs	0.494	0.126	47
InAsSb NWs	0.380	0.133	42
InAsSb/InAs MQWs	0.484	0.099	97
InAs bulk	0.417	0.276	93
InSb bulk	0.235	0.320	170

An Arrhenius plot is shown in Figure 11 for the InAsSb MQW NWs from which an activation energy of 49 meV was obtained from the high-temperature region and ~5 meV for the low-temperature region. This is in approximate agreement with the confinement energy for thermal excitation of holes out of the QW (hole localisation energy of 34 meV) and electrons from the interface triangular QWs (<13 meV localisation energy), respectively. The activation energy of ~49 meV obtained from the Arrhenius plot is consistent with quenching due to carriers escaping confinement rather than Auger recombination and so provides indirect evidence for Auger suppression.

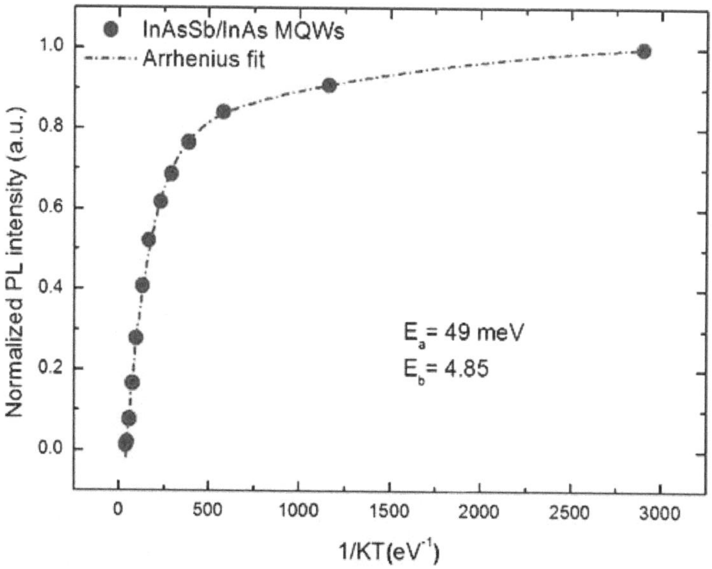

Figure 11. Temperature-dependent analysis of PL emission intensity from the InAsSb/InAs MQWs sample shows an Arrhenius plot of integrated PL intensity as a function of inverse temperature for the InAsSb/InAs MQW NWs. The curve was used to extract activation energies for the mechanisms that drive thermal quenching. Figure obtained with permission from authors [38].

In type II QWs, the Auger rate is determined by the overlap integral (between initial and final electron states at small transferred momentum) which is a minimum when the valence band offset is about three times the conduction band offset. Meanwhile, the radiative rate does not depend on the final state of any excited carrier, since it is a two-body process, so the radiative rate can remain comparable with that in a type I QW systems [55].

8. Conclusion

The development of quantum structures in NWs systems has shown the potential to extend the concept of band-gap engineering to optimise the building blocks of such systems, as well as allowing integrating these systems into the leading CMOS technology, providing a promising future for nanotechnology in optics and electronics. This chapter has presented InAsSb MQWs heterostructures within InAs NWs which exhibit mid-infrared emission at room temperature. The type II QWs provided quantum confinement and spatial localisation of the carriers combined with a suppression of Auger recombination, resulting in enhanced PL emission with respect to the bulk InAs NWs. Furthermore, having characterised the effect of charging on the type II QW, the flat-band transition energy was found to be in good agreement with calculations for both samples. These new quantum-structured NWs will allow novel nano-photonic and quantum light sources to be developed for the technologically important mid-infrared spectral range. They can exploit both the general advantages of site-controlled NWs, such as integration with silicon substrates, and also enhance light-matter coupling based on their dimensions and geometry, opening the way for a wide range of applications.

References

1. Tomioka K, Tanaka T, Hara S, Hiruma K, Fukui T. III-V nanowires on Si substrate: Selective-area growth and device applications. IEEE Journal of Selected Topics in Quantum Electronics. 2011;17:1112-1129
2. Kittel C. Introduction to Solid State Physics. 6th ed. New York, USA: John Wiley & Sons; 2004
3. Pankove JI. Optical Processes in Semiconductors. Dover Books on Physics. New York: Dover Publications; 1971
4. Hafaiedh A, Bouarissa N. Quantum confinement effects on energy GaPs and electron and hole effective masses of quantum well AlN. Physica E: Low-dimensional Systems and Nanostructures. 2011;43:1638-1641
5. Joyce H, Docherty C, Gao Q, Tan H, Jagadish C, Lloyd-Hughes J, Herz L, Johnston M. Electronic properties of GaAs, InAs and InP nanowires studied by terahertz spectroscopy. Nanotechnology. 2013;24:214006
6. Fukuda M. Optical semiconductor devices. New York: John Wiley & Sons; 1999. p. 7

7. Yariv A, Yeh P. Photonics: Optical Electronics in Modern Communications. 6th ed. Oxford University Press; 2007. p. 678
8. Reed GT, Knights AP. Silicon Photonics: An Introduction. Chichester, West Sussex, England: John Wiley & Sons Ltd; 2005
9. Schubert EF. Physical Foundations of Solid-State Devices. Troy, New York: Rensselaer Polytechnic Institute; 2009
10. Koblmüller G, Vizbaras K, Hertenberger S, Bolte S, Rudolph D, Becker J, Döblinger M, Amann M, Finley J, Abstreiter G. Diameter dependent optical emission properties of In: As nanowires grown on Si. Applied Physics Letters. 2012;101:053103
11. Naureen S, Sanatinia R, Shahid N, Anand S. High optical quality InP-based nanopillars fabricated by a top-down approach. Nano Letters. 2011;11:4805-4811
12. Thelander C, Björk M, Larsson M, Hansen A, Wallenberg L, Samuelson L. Electron transport in InAs nanowires and heterostructure nanowire devices. Solid State Communications. 2004;131:573-579
13. Wang B, Leu PW. Tunable and selective resonant absorption in vertical nanowires. Optics Letters. 2012;37(18):3756-3758
14. Yi G. Semiconductor Nanostructures for Optoelectronic Devices. Berlin: Springer Berlin; 2014
15. Foldyna M, Yu L, Roca i, Cabarrocas P. Theoretical shortcircuit current density for different geometries and organizations of silicon nanowires in solar cells. Solar Energy Materials & Solar Cells. 2013;117:645-651
16. Du J, Liang D, Tang H, Gao X. InAs nanowire transistors as gas sensor and the response mechanism. Nano Letters. 2009;9:4348-4351
17. Miao J, Hu W, Guo N, Lu Z, Zou X, Liao L, Shi S, Chen P, Fan Z, Ho J, et al. Single InAs nanowire room-temperature near-infrared photodetectors. ACS Nano. 2014;8:3628-3635
18. Hobbs R, Petkov N, Holmes J. Semiconductor nanowire fabrication by bottom-up and top-down paradigms. Chemistry of Materials. 2012;24:1975-1991
19. Wang H, Sun M, Ding K, Hill M, Ning C. A top-down approach to fabrication of high quality vertical heterostructure nanowire arrays. Nano Letters. 2011;11:1646-1650
20. Koto M. Thermodynamics and kinetics of the growth mechanism of vapor–liquid–solid grown nanowires. Journal of Crystal Growth. 2015;424:49-54
21. Mousavi S, Haratizadeh H, Minaee H. Comparison of structural and photoluminescence properties of zinc oxide nanowires grown by vapor–solid and vapor–liquid–solid methods. Thin Solid Films. 2012;520:4642-4645
22. Obi D, Nechache R, Harnagea C, Rosei F. Mechanical and electrical properties of epitaxial Si nanowires grown by pulsed laser deposition. Journal of Physics: Condensed Matter. 2012;24:445008

23. Radhakrishnan G, Freundlich A, Charlson J, Fuhrmann B. III-V semiconductor vertical and tilted nanowires on silicon using chemical beam epitaxy. MRS Proceedings. 2007;1031:H13-01
24. Dayeh SA, Yu ET, Wang D. Growth of InAs nanowires on SiO2 substrates: Nucleation, evolution, and the role of Au nanoparticles. Journal of Physical Chemistry C. 2007;111:13331
25. Tchernycheva M, Travers L, Patriarche G, Glas F, Harmand JC, Cirlin GE, Dubrovskii VG. Au-assisted molecular beam epitaxy of InAs nanowires: Growth and theoretical analysis. Journal of Applied Physics. 2007;102:094313
26. Zhang X, Zou J, Paladugu M, Guo Y, Wang Y, Kim Y, Joyce HJ, Gao Q, Tan HH, Jagadish C. Evolution of epitaxial InAs nanowires on GaAs (111) B. Small. 2009;5:366-369
27. Ihn SG, Song JI. InAs nanowires on Si substrates grown by solid source molecular beam epitaxy. Nanotechnology. 2007;18:355603
28. Henini M. Molecular Beam Epitaxy. Burlington: Elsevier Science; 2013
29. Ashok S. Defect and Impurity Engineered Semiconductors and Devices III. Warrendale, Pa: Materials Research Society; 2002
30. Sun M, Leong E, Chin A, Ning C, Cirlin G, Samsonenko Y, Dubrovskii V, Chuang L, Chang-Hasnain C. Photoluminescence properties of InAs nanowires grown on GaAs and Si substrates. Nanotechnology. 2010;21:335705
31. Trägårdh J, Persson A, Wagner J, Hessman D, Samuelson L. Measurements of the band gap of Wurtzite InAs1–XPX nanowires using photocurrent spectroscopy. Journal of Applied Physics. 2007;101:123701
32. Möller M, de Lima M Jr, Cantarero A, Chiaramonte T, Cotta M, Iikawa F. Optical emission of InAs nanowires. Nanotechnology. 2012;23:375704
33. Rota M, Ameruddin A, Fonseka H, Gao Q, Mura F, Polimeni A, Miriametro A, Tan H, Jagadish C, Capizzi M. Bandgap energy of wurtzite InAs nanowires. Nano Letters. 2016;16:5197-5203
34. Treu J, Bormann M, Schmeiduch H, Döblinger M, Morkötter S, Matich S, Wiecha P, Saller K, Mayer B, Bichler M, et al. Enhanced luminescence properties of InAs–InAsP core–shell nanowires. Nano Letters. 2013;13:6070-6077
35. Jurczak P, Zhang Y, Wu J, Sanchez A, Aagesen M, Liu H. Ten-fold enhancement of InAs nanowire photoluminescence emission with an InP passivation layer. Nano Letters. 2017;17:3629-3633
36. Sourribes MJL, Isakov I, Panfilova M, Liu H, Warburton PA. Mobility enhancement by Sb-mediated minimisation of stacking fault density in InAs nanowires grown on silicon. Nano Letters. 2014;14:1643-1650. PMID: 24502770
37. Brotherton SD, Lowther JE. Electron and hole capture at Au and Pt centers in silicon. Physical Review Letters. 1980;44:606-609
38. Alhodaib A, Noori Y, Carrington P, Sanchez A, Thompson M, Young R, Krier A, Marshall A. Room-temperature mid-infrared emission from faceted InAsSb multi quantum wells embedded in InAs nanowires. Nano Letters. 2017;18:235-240

39. Johansson J, Bolinsson J, Ek M, Caroff P, Dick KA. Combinatorial approaches to understanding polytypism in IIIV nanowires. ACS Nano. 2012;6:6142-6149. PMID: 22681568

40. Dick K, Thelander C, Samuelson L, Caroff P. Crystal phase engineering in single InAs nanowires. Nano Letters. 2010;10:3494-3499

41. Wu J, Ramsay A, Sanchez A, Zhang Y, Kim D, Brossard F, Hu X, Benamara M, Ware M, Mazur Y, et al. Defect-free self-catalyzed GaAs/GaAsP nanowire quantum dots grown on silicon substrate. Nano Letters. 2015;16:504-511

42. Reimer M, Bulgarini G, Akopian N, Hocevar M, Bavinck M, Verheijen M, Bakkers E, Kouwenhoven L, Zwiller V. Bright single-photon sources in bottom-up tailored nanowires. Nature Communications. 2012;3:737

43. Yang L, Motohisa J, Takeda J, Tomioka K, Fukui T. Size-dependent photoluminescence of hexagonal nanopillars with single InGaAs/GaAs quantum wells fabricated by selective-area metal organic vapor phase epitaxy. Applied Physics Letters. 2006;89:203110

44. Ra Y, Navamathavan R, Park J, Lee C. High-quality uniaxial InxGa1–XN/GaN multiple quantum well (MQW) nanowires (NWs) on Si (111) grown by metal-organic chemical vapor deposition (MOCVD) and light-emitting diode (LED) fabrication. ACS Applied Materials & Interfaces. 2013;5:2111-2117

45. Armitage R, Tsubaki K. Multicolour luminescence from InGaN quantum wells grown over GaN nanowire arrays by molecular-beam epitaxy. Nanotechnology. 2010;21:195202

46. Farrell A, Lee W, Senanayake P, Haddad M, Prikhodko S, Huffaker D. High-quality InAsSb nanowires grown by catalyst-free selective-area metal–organic chemical vapor deposition. Nano Letters. 2015;15:6614-6619

47. Thompson M, Alhodaib A, Craig A, Robson A, Aziz A, Krier A, Svensson J, Wernersson L, Sanchez A, Marshall A. Low leakage-current InAsSb nanowire photodetectors on silicon. Nano Letters. 2016;16:182-187

48. Carrington P, Zhuang Q, Yin M, Krier A. Temperature dependence of mid-infrared electroluminescence in type II InAsSb/InAs multi-quantum well light-emitting diodes. Semiconductor Science and Technology. 2009;24:075001

49. Lord SD. NASA Tech. Memo. 103957. 1992

50. Svensson J, Anttu N, Vainorius N, Borg B, Wernersson L. Diameter-dependent photocurrent in InAsSb nanowire infrared photodetectors. Nano Letters. 2013;13:1380-1385

51. Heber JD, Philips CC. InAs/InAsSb emitters for the mid-infrared region. In: Cai WZ. III-V Semiconductor Heterostructures: Physics and Devices. Keraia: Research Signpost; 2003. pp. 140-167

52. De A, Craig E. Predicted band structures of III-V semiconductors in the wurtzite phase. Physical Review B. 2011;84:239907

53. Olsson L, Andersson C, Håkansson M, Kanski J, Ilver L, Karlsson U. Charge accumulation at InAs surfaces. Physical Review Letters. 1996;76:3626-3629

54. Latkowska M et al. Temperature dependence of photoluminescence from InNAsSb layers: The role of localized and free carrier emission in determination of temperature dependence of energy gap. Applied Physics Letters. 2013;102:122109

55. Zegrya GG, Andreev AD. Mechanism of suppression of Auger recombination processes in type II heterostructures. Applied Physics Letters. 1995;67(18):2681

6

Parasitic Capacitances on Scaling Lateral Nanowire

1. Introduction

Since the beginning of solid-state technology, a continuous reduction of transistor size has delivered the smaller and faster electronic technology in every new generation. Since the last few decades, Moore's law [1] was gifted with many prime movers, such as the mobility boosted by strained silicon [2] and reduced gate leakage using the high-k/metal gate [3] and most importantly, planar MOSFET was replaced by the FinFET to achieve better leakage control [4]. Since then, the major industries followed FinFET technology for their 16/14 nm and 10 nm node, respectively [5, 6, 7]. However, while scaling down below the 10 nm node, again the short channel effects (SCE) such as subthreshold leakage rise significantly and became a major concern in scaling FinFET architecture. As per international technology roadmap for semiconductor (ITRS 2015) prediction, the contacted gate pitch (CGP) in 7 nm node transistor would be ~42 nm, making a gate length of less than 15 nm [8]. Although the FinFET has gate wrapping around the channel, at these shorter dimensions, a much stronger gate control is required. Therefore, gate-all-around (GAA) architecture has emerged as an alternative to FinFET [9]. The gate in NW covered all over around silicon channel providing a stronger control, therefore preventing more unwanted leakages. However, reduction of effective width (Weff) in NW reduces the current driving capacity significantly. Although the drive current in NW can be increased by stacking multiple wires per fin, however, a taller fin device increases parasitic capacitances which may limit the benefit of scaling [10]. Though numerous studies have been carried out for analyzing the intrinsic and parasitic capacitance [11, 12, 13, 14], however, there is still a requirement for an extensive analysis to model the GAA-NW's major parasitic components. Thus, this chapter deals with the capacitance model of a GAA-NW transistor as well as the overall scaling performances at ring oscillator circuits.

2. GAA-NW device

To continue Moore's law, transistor sizes are scaled down to the 7 nm node (N7) and 5 nm node (N5) specifications [15, 16]. A contacted gate pitch (CGP) of 42 and 32 nm were used in both the N7 and N5 devices. Gate length (Lg) of 14 and 10 nm with a wire diameter of 7 (D7) and 5 nm (D5) was considered in all the N7 and N5 specifications [10]. Channel material for n-channel and p-channel GAA-NW was considered with Si and Si50Ge50, respectively. An epitaxial-shaped source (S) and drain (D) regions were used as contacts. The S/D regions were doped with an active doping concentration of 3 × 1020/cm3 for both n-channel and p-channel devices with a specific contact resistivity of 5 × 10−9 Ωcm2. Gate dielectric of 0.5 nm oxide and 1.5 nm high-k (Hf02) layers was used. A gate spacer of 5 nm thickness was applied to both the gate-to-source and gate-to-drain regions (relative permittivity, εr = 4.4). A midpoint work function value for metal gate was used in both the devices during initial simulation. Considering a fixed NW pitch (NWP) of 14 and 10 nm offset, the two-stacked-NW device creates a fin height of 31 and 29 nm, respectively. Other specifications and setup for each device were considered similar to the reference [10, 15]. The simulated compact NW is shown in Figure 1(a) and a cross-sectional view showing inside details in Figure 1(b).

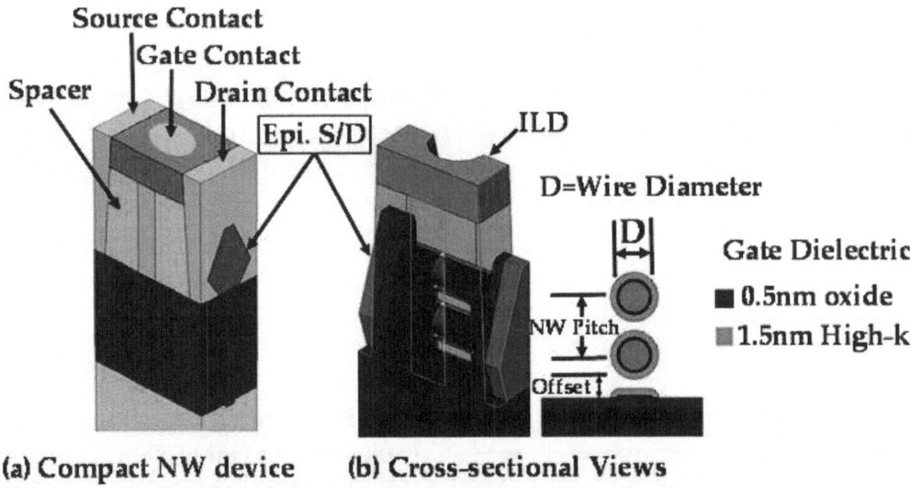

Figure 1. Compact GAA-nanowire transistor with cross-sectional views. (a) Compact NW device. (b) Cross-sectional views.

3. Simulation methodology

To estimate various parasitic capacitances in a GAA-NW and their impacts on overall scaling performances at a higher frequency, a TCAD-based compact model study was performed. Figure 2 shows a flow diagram used to analyze the DC and AC performances of the scaled GAA-NW transistor.

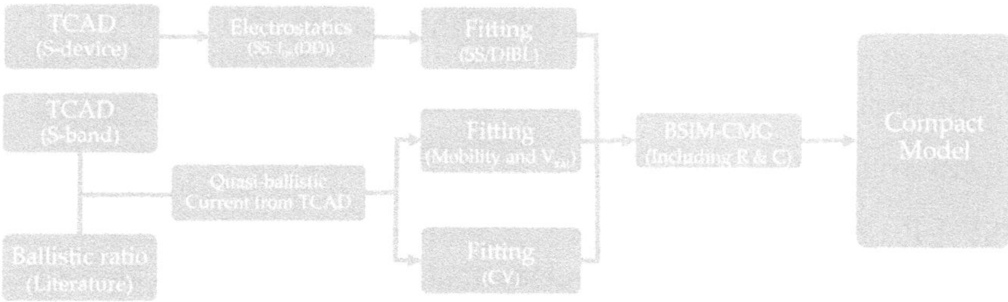

Figure 2. Flow diagram describes the TACD-based compact model of GAA-NW.

A two-stacked GAA-NW transistor was implemented using the Sprocess module of TCAD tool Sentaurus [17]. Given a continuous channel in FinFET, the selective region was etched away to form a round-shaped NW channel [18]. The 7 nm width-based FinFET transformed into the 7 nm diameter-based GAA-NW. Then the diameter and the gate length of NW were scaled down further to 5 and 10 nm, respectively [10]. All the parameters for device simulation were considered similar to the default 7 nm FinFET model [17].

3.1. Subthreshold current estimation

Electrical performances were estimated separately for both the off-state and for the on-state conditions. The Sdevice [19] simulation with Shockley-Read-Hall (SRH), auger, band-to-band tunneling (BTBT) recombination, bandgap narrowing, anisotropic density gradient, interface charge, mobility model with multivalley correction, thin inversion layer correction with high-k dielectric and quantum correction for the inversion layer were used to obtain the properties such as subthreshold slope (SS) and drain-induced barrier lowering (DIBL). Then, these intrinsic properties (SS and DIBL) for different gate lengths, wire diameters and vertical pitches were fitted into the BSIM-CMG model [20]. Thus the fitted model provided subthreshold characteristics of a NW used for circuit-level simulations.

3.2. The drive current estimation

To estimate the on-state drive current, the ballistic flow has been considered along with drift-diffusion currents. As the channel length in N7 and N5 devices was scaled down to the carrier's mean free path range [21], the total drive current in NW was considered to be quasi-ballistic in nature [21]. The scattering-free ideal ballistic current is defined by Eq. (1), where q is the electronic charge, Vinj is the carrier's injection velocity and Ninv is the number of inversion charges. Sband [22] simulation provided these pure ballistic current characteristics for the different wire diameter and applied voltages

I Bal = q × V inj × N inv E1

The quasi-ballistic current is represented by Eq. (2). Quasi-ballistic current is the product of pure ballistic current times the ballistic ratio. The ballistic ratio (BR) is the ratio between actual saturation current and ideal ballistic currents.

I quasi – ballistic = I Bal × BR E2

The ballistic current is independent of gate lengths, but BR is strongly dependent on the gate length, wire diameter and channel stress [21]. The BR for D7 NW was assumed to be similar to a 7 nm width-based FinFET although the NW might have lower BR than FinFET; since it strongly depends on the body configuration of the device, this possible small error on BR was further screened by electrostatics and access resistance [10]. The BR for D5 device (Figure 3) was extrapolated because it is expected to have significantly lower BR than the D7, similar to [10]. The variations of BR with the applied channel stress and channel length reduction for both the n-channel and p-channel FinFET (tensile for NMOS and compressive for PMOS) are shown in Figure 4(a) and (b) [23, 24, 25, 26].

$$BR = \frac{\textit{Actual Saturation Current}}{\textit{100\% Ballistic Current}}$$

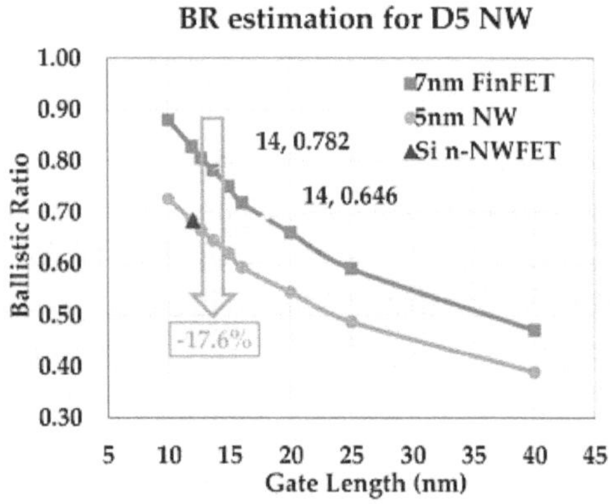

Figure 3. Ballistic ration (BR) considered for the 7 nm diameter (D7) based of LNW and extrapolated BR for the 5 nm diameter (D5) GAA-NW.

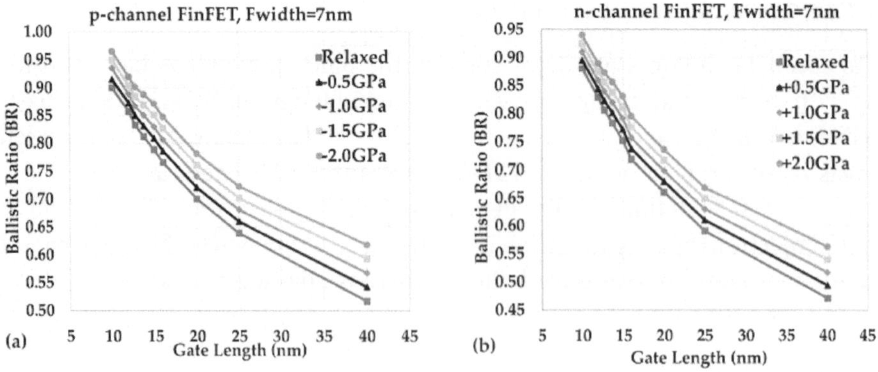

Figure 4. BR variation on varying channel stress in both the n-channel and p-channel FinFET's.

Next, for both the D5 and D7 devices, Sband simulated results such as carrier's injection velocity (Vinj) and the number of inversion charges (Ninv) were multiplied with electronic charge (q) to obtain the IBal, as plotted in Figure 5(a) [10].

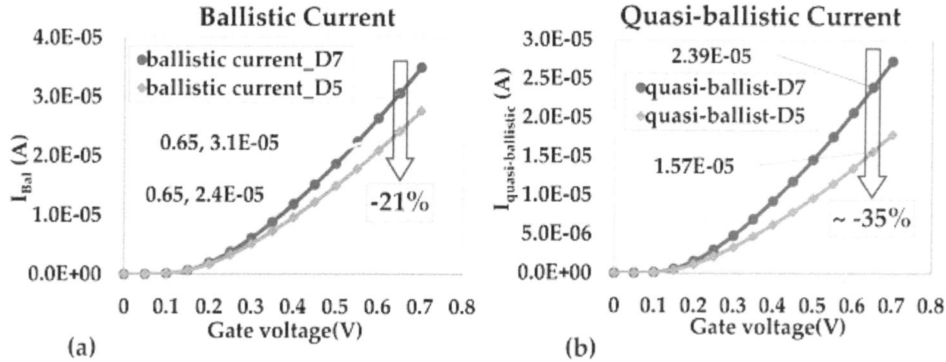

Figure 5. Ballistic characteristics: (a) simulated ideal ballistic currents for both the D7 and D5 n-channel GAA-NW, (b) quasi-ballistic currents considering the extracted BR.

The variations of pure ballistic current (IBal) with the gate voltages are plotted in Figure 5(a) and after multiplying with BR, the obtained Iquasi-ball is shown in Figure 5(b). Then, these quasi-ballistic characteristics were fitted into the BSIM-CMG model by fitting the mobility and carrier's velocity equations similar to [10]. As well as the intrinsic capacitances, interface trap and other device properties for each NW were fitted into the BSIM-CMG model.

The final drive current was obtained after including the front end of the line/ mid of the line (FEOL/MOL) R&C parasitics into the BSIM-CMG model file [20]. Then, SPICE simulation was performed with this fitted model to obtain high-frequency properties for a 15th-stage ring oscillator (RO) inverter circuits [27]. The ring oscillator setup is discussed in Section 6. All the simulations were performed at a fixed saturation drive voltage of 0.65 V and targeted off-state current (Ioff) of 3.5 nA per fin.

4. Parasitic capacitance estimation

In this section, various parasitic capacitances associated with the metal gate of the NW transistors are discussed. We start with a basic fin channel architecture and then the formation of vertical nanosheet and then finally a GAA-NW with a higher vertical pitch. The vertical stacking of GAA-NW is essential for higher drive currents, but then again, the area associated with the wrapping metal lines is also increased as the fin height is increased, thus by doing so expected stronger parasitic capacitances.

4.1. Vertical pitch variation (fin to NW)

The transition from fin channel to GAA channel enables much stronger gate control. The bottom region of the channel fin was released and covered with the gate dielectric layers, as shown in Figure 6(I), as a result of making the channel gate-all-around vertical nanosheet. Next, the single nanosheet was gradually transformed into two isolated NWs. Subsequently the vertical pitch between two NWs was increased from 7 to 19 nm (Figure 6(III)–(IX)). The transition from a single NS to multiple NW improves the subthreshold slope (SS) characteristic due to a stronger control in all-around architecture. The variations of subthreshold slope (SS) for all these structures have been plotted in Figure 7(a). The SS values are improved by shifting from Figure 6(I)–(V) but remain constant after Figure 6(VI)–(IX). This is because, once a uniform metal layer is processed between two NWs, the work function fluctuations might be stabilized, thus providing a maximum SS improvement. However, the variations of active channel area in Figure 6(I)–(IV) also affect the drive currents. The deviation of Ion (drift-diffusion) for all the structures was plotted in Figure 7(b). Thus it can be concluded that up to a certain fin height with higher vertical pitch, NW stacking improves SS compared to a continuous fin channel; however, SS benefit may be limited by the lowered active channel area along with the increased parasitic capacitances.

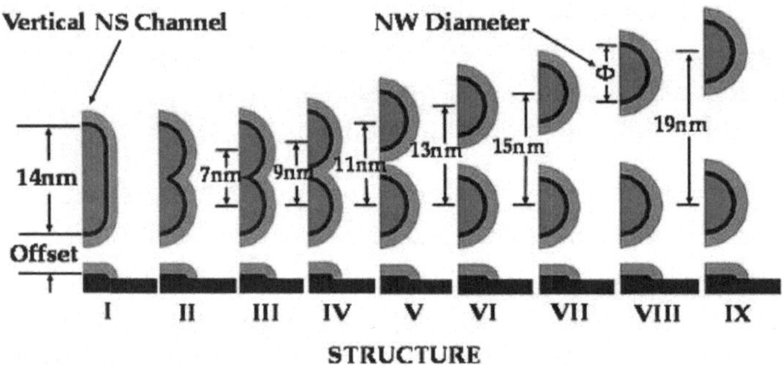

Figure 6. The channel regions cross-sectional view for GAA transistors: Fin to vertical nanosheet (NS) to higher pitch nanowires (NW).

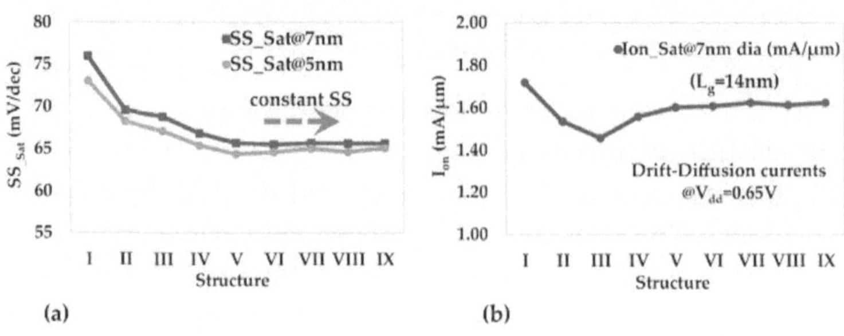

Figure 7. Electrostatics for all the structures shown in Figure 6: (a) sub-threshold slope (SS) variations, (b) change in currents flows due to the change in channel area.

4.2. The capacitance model

To analyze the overall parasitic capacitances on increasing the fin height, a schematic model depicting gate and source contact lines of a NW is shown in Figure 8. This model represents half of the NW architecture as a simplified model for calculating major capacitances associated with gate-to-source sidewall only. Similarly, the gate-to-drain sidewall capacitances can be calculated by reflecting this model as gate and drain contact sidewalls. At this point, the model represents four different major capacitances Cp1, Cp2, Cp3 and Cp4. The capacitance Cp1 represents the wire fringe capacitances between the gate sidewall and all the nanowire surfaces. The spacer between the source contact and gate contact line was considered to be (Tspacer) 5 nm; this narrow spacing forms a strong parallel palate capacitance between gate and source. Therefore, the Cp2 represents the major parasitic capacitance between source sidewall and gate sidewall. The Cp2 is expected strongly dependent on NW's vertical pitch. The Cp3 is a fringe capacitance between the top gate surface and the source contact sidewall. And, Cp4 is an overlap (around 2 nm) capacitance between the nanowire and gate metal line. As we have seen previously, achieving a better electrostatic control with higher drive performance, the multiple stacking with higher vertical pitches are essential. However, both of these requirements increase the fin height as well as the Cp2. These two conflicting points need to be carefully optimized in both the process variation and for the best electrical performances. Among these capacitances, the Cp2 has a major contribution on overall device performance; thus, a detailed calculation methodology of Cp2 is presented here.

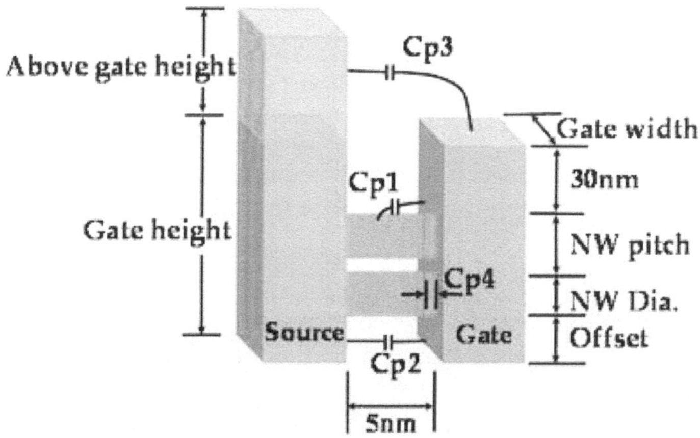

Cp1=Fringe capacitance; Cp2=Parallel plate capacitance

Cp3=Fringe capacitance between the gate top and the S/D contact

Cp4=Overlap capacitance ($L_{overlap}$=2nm)

Figure 8. The capacitance model representing various parasitic capacitances.

4.3. Parallel plate capacitance (Cp2)

In calculating the values of Cp2, a conventional parallel plate capacitance formula (εA/d) was used, wherein d and ε are the spacer width and permittivity, respectively. However, the active area (A) was calculated by considering a special development. Area (A) is the average of AREAsource and AREAgate wherein the AREAsource is the cross-marked source sidewall region after excluding the wire diameters as shown in Figure 9. And AREAgate is the cross-marked gate sidewall regions after excluding wire regions as well as by drawing a new perimeter related to the spacer width (Tspacer = 5 nm). This is mainly to exclude the contributions of continuous electric flux shared by two parallel plates through the connected NWs. The NW with two different wire diameters (7 nm = D7 and 5 nm = D5) was varied for different vertical pitches. Values of all the capacitances are calculated and presented in the next section.

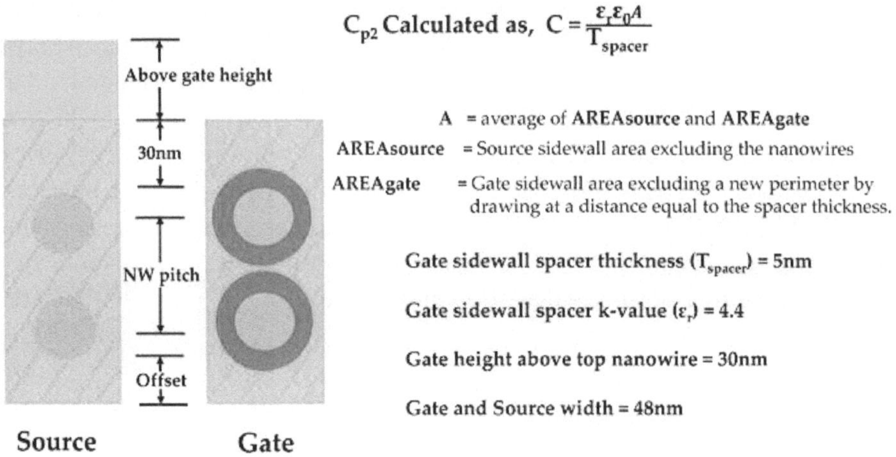

Figure 9. The cross-sectional side view of source and gate contacts regions for calculating the active area in Cp2.

4.4. The capacitance values

The comprehensive model formula presented in papers [11, 12] is used to calculate all the parasitic capacitances. Considering the initial values from [12] the capacitances, Cp1, Cp2, Cp3 and Cp4 are plotted for all the architectures presented in Figure 6. Figure 10(a) shows the Cp1 variations with the vertical pitches. As per the active area shape changes in Figures 6(I)–(IV), a continuous increase of wire fringe capacitance Cp1 was observed; however, it remains constant after Figure 6(V)–(IX) for the higher vertical pitches. The values of parallel plate capacitance Cp2 were calculated after considering the effective area of S/G sidewalls and have been plotted in Figure 10(b). It was observed that, unlike CP1, the values of Cp2 experience a continuous rise with increasing the vertical pitch from Figure 6(I)–(IX). The overlap capacitance Cp3 remains constant as the top surface of gate metal and source contact is similar for all the structures (Figure 11). Cp4 shows increases till Figure 6(IV) and then remains constant for Figure 6(V)–(IX) (similar to the trend of Cp1). Although the overlapped region of 2 nm was considerably minor, the values of Cp4 show marginally higher than the values of fringe capacitance Cp1. Among all these capacitances, Cp2 witnessed the most significant parasitics which is strongly dependent on fin height (vertical pitch). The value of Cp2 in Figure 6(IX) experienced a 22% higher capacitance than in Figure 6(I).

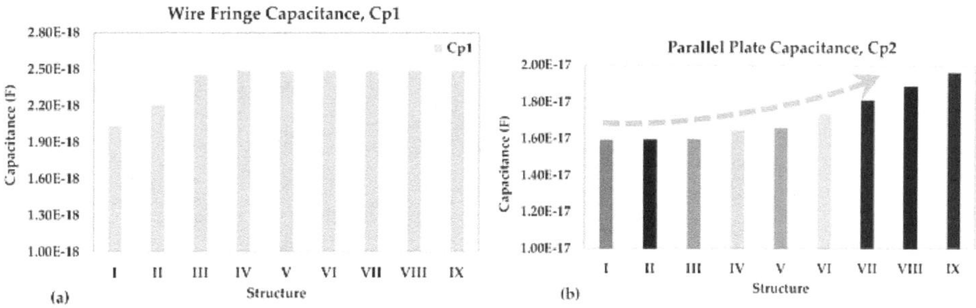

Figure 10. Parasitic capacitance values: (a) the initial increasing trend of Cp1 but remains same after structure IV, (b) Cp2 trends to increase continuously with the vertical pitch.

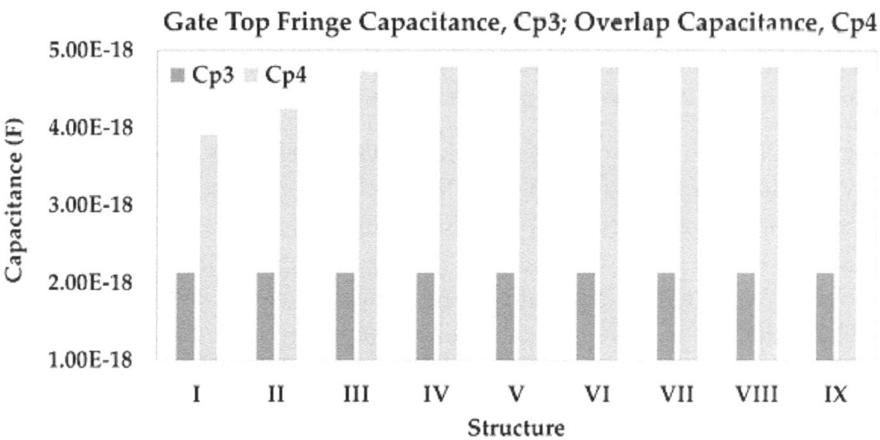

Figure 11. Fringe capacitance Cp3 and overlap capacitance Cp4 plotted in the same axis.

5. Ring oscillator simulation

To benchmark the power performance and area gain for both the 7 nm diameter (D7)- and the 5 nm diameter (D5)-based GAA-NW, ring oscillator-level circuit simulations were performed [10]. Considering an inverter-based ring oscillator (RO), the RC delay and active power consumption as well as leakage power loss were calculated for each device. The ring oscillator (RO) with a 15-stage inverter was simulated using the SPICE simulation setup [27]. All the inverter stages had a fan out three and loaded with an optimum back end of the line (BEOL) load [28]. A median value of interconnect wire length for the critical path was loaded as the BEOL load including all parasitics in each inverter stage (50 CGP long) [28]. Then, a minimum delay optimization technique for a critical path was used to optimize the benefits of scaling D5 NW with various device configurations such as tighter vertical pitches and reduced gate lengths. The schematic of RO chain setup is shown in Figure 12(a).

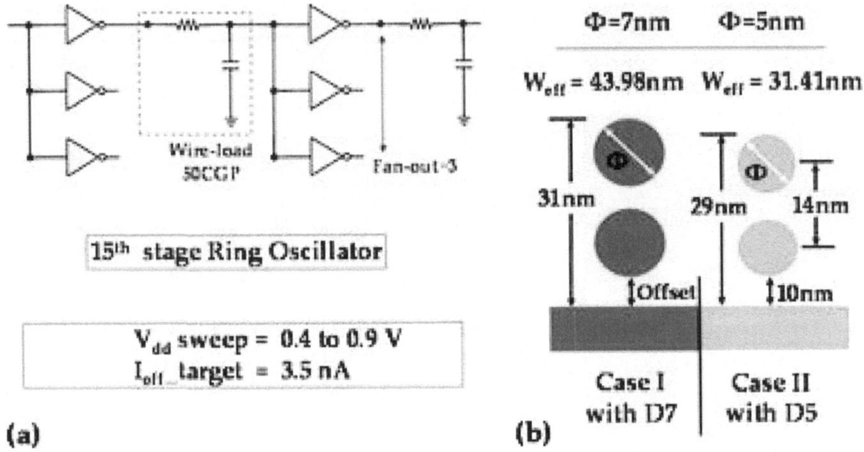

Figure 12. Ring oscillator testbench: (a) the inverter chain with FO three (b) cross-sectional NW showing the wire diameters and various specifications.

Figure 12(b) shows two-stacked D7 and D5 NW devices. Moving from D7 to D5 NW provides a 26% reduction in Weff as well as a 2 nm reduction in fin height (with an equal vertical pitch of 14 nm). Reducing the wire diameter from D7 to D5 offers only 2 mV/decade and 4 mV/V SS and DIBL improvements [10]. However, reduction in Weff delivers a significant reduction in overall dive current (~35%) [10]. This reduction in drive current might be improved by reducing the NW's gate length from 14 to 10 nm as well as by stacking more numbers of NW per fin with a reduced vertical pitch. On the other hand, stacking multiple NW increases parasitic capacitances along with the fin height. In this section, we have discussed the benefit of scaling on power and speed performance for both the 7 nm node (D7/D5 @N7) and the 5 nm node (D5@N5) GAA-NW.

$$P_{dyn} = f C_{eff} V_{dd}^{2} \quad (3)$$

5.1. Power-delay optimization at 7 nm node (N7) dimensions

For all the 7 nm node (N7) specifications, a 42 nm CGP and 14 nm gate length were considered [10]. The dynamic power consumptions in an oscillator circuit were calculated based on the formula presented by Eq. (3), where Ceff is the effective load capacitance, f is the operating frequency and Vdd is the drive voltage. The RC delay was calculated in both the D7 and D5 devices based on the values obtained in [10].

The change in ring oscillator performances (RC delay and power loss) with the variation of Vdd from 0.4 to 0.9 V is displayed in Figure 13. The optimum delay versus drive voltage variation has been presented in Figure 13(a). Figure 13(b) shows the active power loss for both the D7 and the D5 devices. Nearly a 20% rise in delay was observed for the D5 device compared to the D7 device, largely due to the reduction of drive currents. This reduced drive current (active area) in D5 device at the same applied voltage consumes less active power than the D7 device. However, the actual benefit can be visualized by plotting the power-delay product (energy) and leakage power loss for both the devices at a targeted frequency.

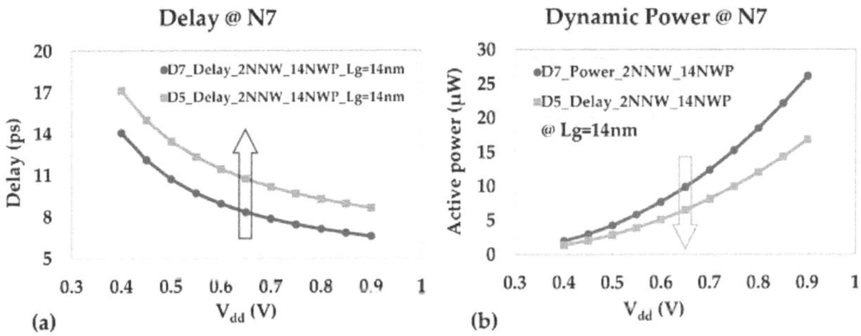

Figure 13. Performance at 42 nm CGP: (a) delay for D7 and D5, (b) power for both D7 and D5 NW.

Figure 14(a) shows a change in energy consumption with the variations of frequency for both the devices. While operating at the same frequency, a significant rise in energy consumption was observed for the D5 device in comparison to the D7 device. To achieve an equal drive current in the D5 device always require a higher drive voltage in comparison to the D7 device, therefore consuming more energy leading to overall degraded performances. Besides that, while operating at the same frequency, the D5 device consumes more leakage power than the D7 device (Figure 14(b)).

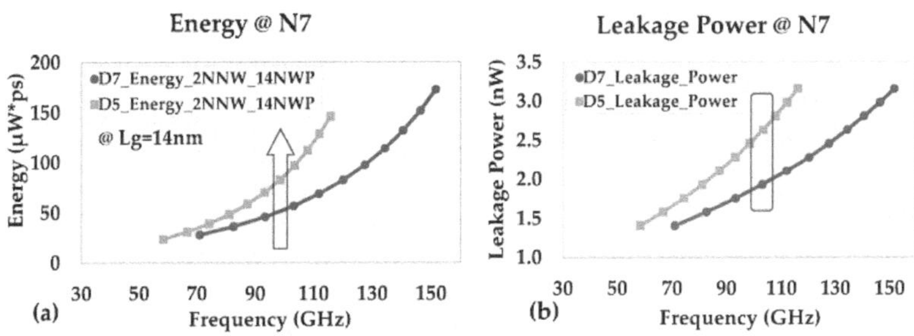

Figure 14. Speed estimation for both the D7 and D5 NW with: (a) energy, (b) leakage power.

5.2. Power-delay optimization at 5 nm node (N5) dimensions

Although the electrostatics were improved marginally by reducing the nanowire's diameter, the overall performance was degraded at the same gate length. Therefore, in order to understand the overall benefit of scaling the wire diameter, we scaled the gate length from 14 to 10 nm. At the 10 nm gate length, the D5 showed a significant SS improvement compared to the D7 device (~10 mV/decade) [10]. A 10 nm gate length and 32 nm CGP were defined in D5 NW for the N5 specifications. Furthermore, the vertical pitch between two D5 NWs was reduced to 12 nm. To equalize the drive current, a four-stacked D5 NW at N5 specifications was compared with the two-stacked D7 NW at N7 specifications with increased BEOL load. In this case, the D5 shows an improved delay performance compared to the D7 device (Figure 15(a)). It should be noted that the D5 consumes more active power than the D7 device, as shown in Figure 15(b). These minimal improvements in delay and power were further normalized by the energy versus speed performances, as shown in Figure 16(a). Figure 16(b) shows the comparable leakage power characteristics for both the devices. Though, after considering taller fin in the D5 case, the power-delay product shows a comparable trend for low-frequency operation, however, it starts consuming more energy as it goes toward high-frequency operation. Now, with these NW configurations, a fin height in four-stacked D5 devices grows an additional 14 nm taller than a two-stacked D7 device adding an additional parasitic capacitance as expected.

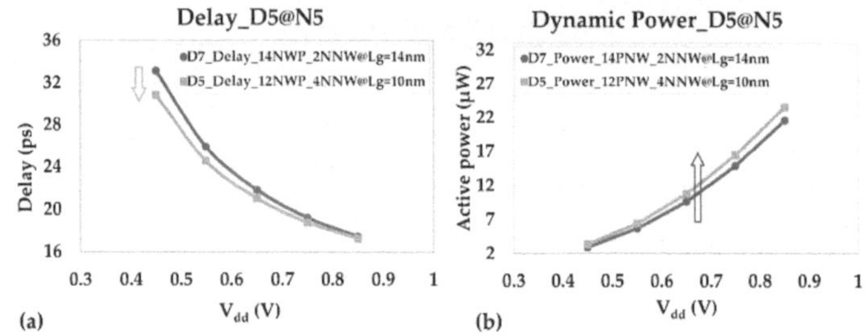

Figure 15. Performances (D5@32 nm and D7@42 nm CGP): (a) delay versus Vdd, (b) power versus Vdd.

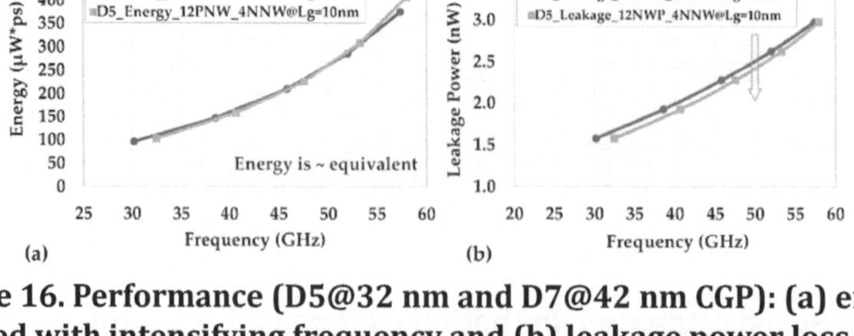

Figure 16. Performance (D5@32 nm and D7@42 nm CGP): (a) energy consumed with intensifying frequency and (b) leakage power loss with the frequency deviations.

This flexibility shows the various aspects of scaling the D5 NW at a reduced wire pitch along with a greater number of NW (NNW) stacking. The overall performance benefits for a D5 device at N5 specifications and D7 device at N7 specifications were benchmarked with the variations in fin height. In the next section, we discussed the energy versus frequency for different NW alignments.

5.3. Benchmarking D5 NW at N5 and D7 NW at N7

A three-stacked D5 device delivers nearly equivalent drive currents with a two-stacked D7 device (only 6% loss) [10]. Subsequently, using D5 NW the drive current can be increased by stacking a greater number of wires per fin as the D5 may provide an area benefit with a 32 nm CGP. On the other hand, stacking more numbers of D5 NW will increase the parasitic capacitance (mainly Cp2) in comparison to the rise in gate capacitances (FEOL). The fin height variation with different NW stacking is shown in Figure 17. When moving from two-stacked to five-stacked NW D5 NW device, a 91% rise in drive current has been observed (18.03–34.44 μA), but this increases the total fin height by 90% (from 29 to 55 nm). On the other hand, moving from two-stacked to three-stacked D7 NW device (14 nm wire pitch) improves the drive currents by 30.5% (25.95–33.87 μA), but the fin height is increased by 45% (from 31 to 45 nm) only. The three-stacked D7 NW delivers almost equal drive currents with the five-stacked D5 NW at much lower fin height (33.87 and 34.44 μA). Thus, after considering all the parasitic impacts, a fair comparison is needed to observe the actual benefit of scaling wire diameter (D5).

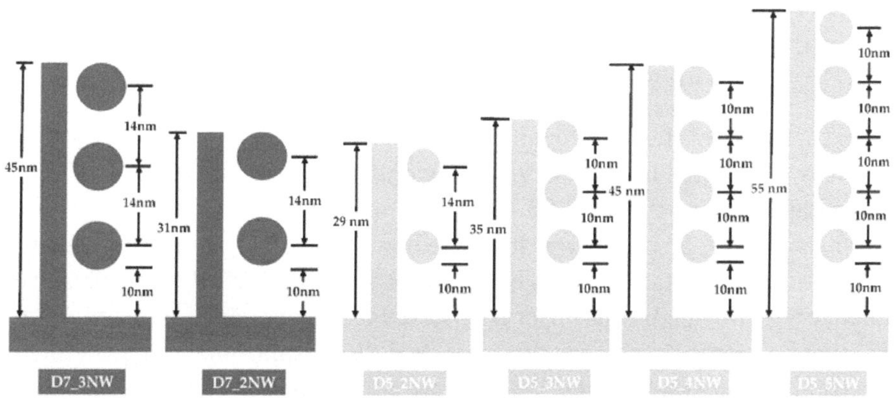

Figure 17. The cross-sectional NWs schematic showing upswing in fin heights with multiple stacking of NWs.

Optimizing the fin height in a NW transistor remains very sensitive to overall device design. Keeping this in mind, the total energy consumptions per device with increased fin height were plotted in Figure 18 for both the D5 and D7 devices with multiple stacking. For both the D7 and D5, a single-stacked NW to a five-stacked NW was considered. Along with the multiple stacking, three different vertical pitches (16, 14 and 12 nm) were varied for the D7 device and four different pitches (16, 14,

12 and 10 nm) for the D5 device. A continuous rise in energy consumptions was observed while going from single NW to multiple NW in both the D7 and D5 devices. The fin height has been reduced by reducing the NW vertical pitch, thus shrinking the energy loss. It was observed that the D7 NW always consumes higher energy than the D5 NW at specific fin height. However, with a lowered fin height, D5 NW delivers lower drive currents which further reduces the overall device speed.

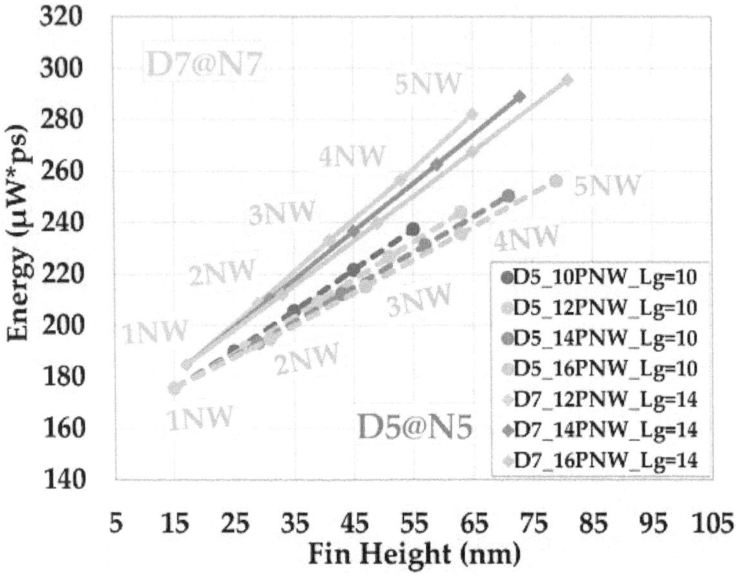

Figure 18. Variations of energy with varying fin height from single stacked to five stacked NW.

The variations of speed (frequency) with increasing the fin height for both D7 and D5 devices are plotted in Figure 19. At an identical fin height, the D7 device always delivers higher speed than the D5 device. Hence, considering both the D7 and D5 NW transistors, the fin height was increased by stacking single wire to five multiple wires. Until a three-stacked design for both the D7 and D5 NWs, a linear rise in frequency with fin height was observed. However, the frequency gets saturated with increasing fin height beyond the three-stacked NW design. Further rise in fin height will start degrading the frequency performance. This is mainly caused by the rise in parasitic capacitances. Generally, the capacitance loaded with a transistor is the sum of both intrinsic gate capacitance and FEOL parasitic capacitances. Increasing the fin height increases the total area of gate metal stack positioned next to a source and a drain contact lines. This effect remains hidden in both the D7 and D5 NWs up to a certain fin height. At the same fin height, the D7 drives have slightly higher speed in comparison to the D5 device as it has lower parasitic influence. Furthermore, increasing the fin height started degrading the total speed performance of D5 device quite early. Hence multiple stacking is also a major limiter for providing a higher speed.

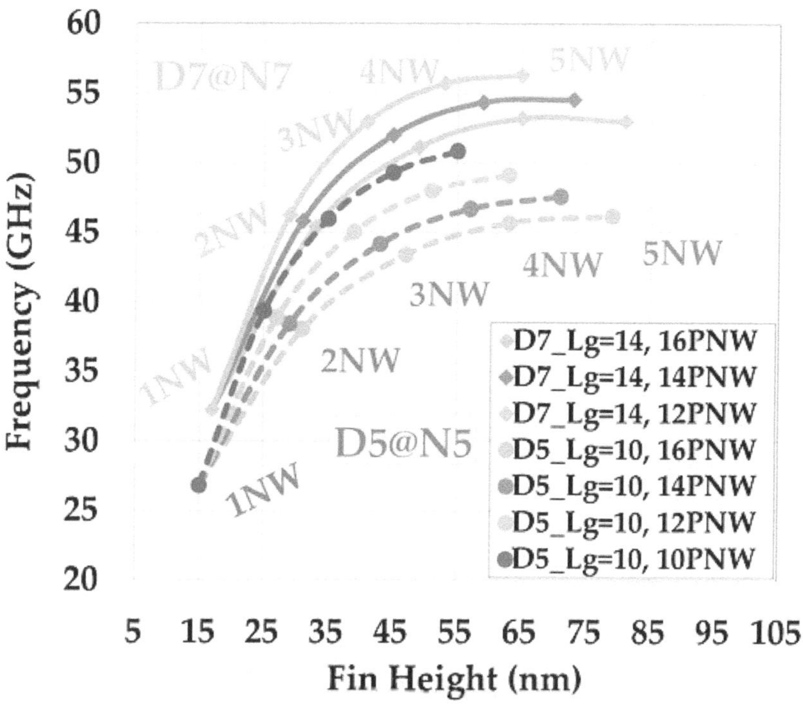

Figure 19. Frequency variations with the fin height from single stacked to five stacked NW.

Along with the speed, the total energy consumption with different frequency operations is also plotted in Figure 20. The energy consumptions are largely affected by the escalation of parasitic capacitances. Due to that reason, it starts consuming more energy while raising the operating frequency. Moving from single NW to five-stacked NW experiences an exponential rise in energy loss in both the devices. Although the single-stacked D5 NW consumes 27% lower energy than the single-stacked D7 NW, it also provides 20% reduction in speed performance. This gap however is quickly normalized while operating at higher frequency. Around 45–50 GHz operation, the D5 device requires four- to five-stacked NW and starts consuming more energy than the two- to three-stacked D7 NW. We have observed that the D5 NW at N5 specification consumes lower energy at the low-frequency operations, but it starts consuming higher energy than the D7 device at high-frequency operation. Multiple stacking increases both on currents as well as parasitic losses. Thus, the NW with shorter fin height should be considered to achieve the benefit of scaling at low-frequency performance.

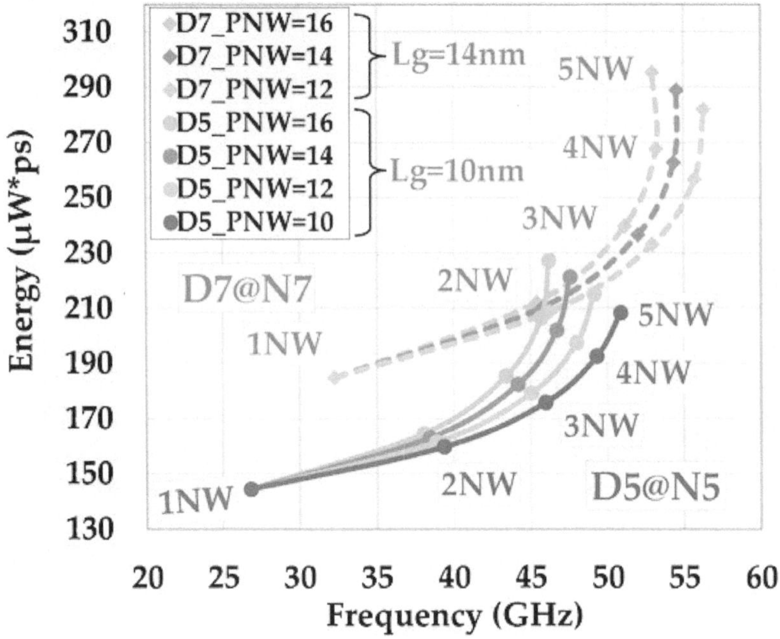

Figure 20. Energy versus frequency variations for both the D7 and D5 NW transistors.

6. Conclusion

The GAA lateral nanowire is a promising candidate for scaling beyond the FinFET technology. It provides superior electrostatic benefits such as better SS and DIBL at the shortest gate length. However, it has a very complex process as well as huge impacts of parasitics. The study conducted here was mainly focused on the effects of all the parasitic capacitances on overall device performance with scaling beyond the 7 nm node dimensions targeting the 5 nm node technology. The TCAD study compared the electrostatic performance and compact modeling which shows a basic inverter operation connected as a ring oscillator. A capacitance model was discussed considering the major parasitic components and their impacts on advance dimensional scaling. The compact modeling showed that reducing the wire diameter improved the electrostatics marginally but degraded the overall performance. Therefore, a deep analytical and experimental study is required to conclude the overall scaling benefits of GAA-NW transistor

References

1. Moore GE. Cramming more components onto integrated circuits. Electronics. 1965;38(8):114-117
2. Mistry K, Armstrong M, Auth C, Cea S, Coan T, Ghani T, Hoffmann T, et al. Delaying forever: Uniaxial strained silicon transistors in a 90 nm CMOS technology. In: Symposium on VLSI Technology; 2004. pp. 50-51

3. Auth C, Cappellani A, Chun JS, Dalis A, Davis A, Ghani T, Glass G, et al. 45 nm high-k+ metal gate strain-enhanced transistors. In: 2008 Symposium on VLSI Technology. IEEE; 2008. pp. 128-129

4. Auth C, Allen C, Blattner A, Bergstrom D, Brazier M, Bost M, Buehler M, et al. A 22 nm high performance and low-power CMOS technology featuring fully-depleted tri-gate transistors, self-aligned contacts and high density MIM capacitors. In: 2012 Symposium on VLSI Technology (VLSIT). IEEE; 2012. pp. 131-132

5. Wu SY, Lin CY, Chiang MC, Liaw JJ, Cheng JY, Yang SH, Chang SZ, et al. An enhanced 16 nm CMOS technology featuring 2nd generation FinFET transistors and advanced Cu/low-k interconnect for low power and high performance applications. In: 2014 IEEE International Electron Devices Meeting (IEDM). IEEE; 2014. pp. 3-1

6. Natarajan S, Agostinelli M, Akbar S, Bost M, Bowonder A, Chikarmane V, Chouksey S, et al. A 14 nm logic technology featuring 2nd-generation FinFET, air-gapped interconnects, self-aligned double patterning and a 0.0588 μm2 SRAM cell size. In: 2014 IEEE International Electron Devices Meeting (IEDM); 2014. pp. 3-7

7. Auth C, Aliyarukunju A, Asoro M, Bergstrom D, Bhagwat V, Birdsall J, Bisnik N, et al. A 10 nm high performance and low-power CMOS technology featuring 3rd generation FinFET transistors, self-aligned quad patterning, contact over active gate and cobalt local interconnects. In: 2017 IEEE International Electron Devices Meeting (IEDM). IEEE; 2017. pp. 29-1

8. International Technology Roadmap for Semiconductors (ITRS) 2.0. The Semiconductor Industry Association, Technical Report [Online]. 2015. Available from: http://www.itrs2.net/itrs-reports.html

9. Kim SD, Guillorn M, Lauer I, Oldiges P, Hook T, Na MH. Performance trade-offs in FinFET and gate-all-around device architectures for 7 nm-node and beyond. In: IEEE SOI-3D-Subthreshold Microelectronics Technology Unified Conference (S3S). IEEE; 2015. pp. 1-3

10. Das UK, Garcia Bardon M, Jang D, Eneman G, Schuddinck P, Yakimets D, et al. Limitations on lateral nanowire scaling beyond 7-nm node. IEEE Electron Device Letters. 2017;38(1):9-11

11. Lacord J, Ghibaudo G, Boeuf F. Comprehensive and accurate parasitic capacitance models for two-and three-dimensional CMOS device structures. IEEE Transactions on Electron Devices. 2012;59(5):1332-1344

12. Lacord J, Martinie S, Rozeau O, Jaud M-A, Barraud S, Barbé JC. Parasitic capacitance analytical model for sub-7-nm multigate devices. IEEE Transactions on Electron Devices. 2016;63(2):781-786

13. Seo Y, Kang M, Shin H. Analysis of parasitic capacitance and performance in gate-al-around and tri-gate channel vertical FET. In: 2017 Silicon Nanoelectronics Workshop (SNW). IEEE; 2017. pp. 63-64

14. Kim J, Woo C, Kang M, Shin H. Analysis on extension region in nanowire FET considering RC delay and electrical characteristics. In: 2017 Silicon Nanoelectronics Workshop (SNW). IEEE; 2017. pp. 43-44
15. Liebmann L, Zeng J, Zhu X, Yuan L, Bouche G, Kye J. Overcoming scaling barriers through design technology co-optimization. In: 2016 IEEE Symposium on VLSI Technology. IEEE; 2016. pp. 1-2
16. Mertens H, Ritzenthaler R, Pena V, Santoro G, Kenis K, Schulze A, Litta ED, et al. Vertically stacked gate-all-around Si nanowire transistors: Key process optimizations and ring oscillator demonstration. In: 2017 IEEE International Electron Devices Meeting (IEDM). IEEE; 2017. pp. 37-4
17. Sentaurus™ Process User Guide, L-2016.12. Mountain View, CA, USA: Synopsys; 2016
18. Sentaurus™ Project Demonstrates. Three-dimensional device simulations of 10 nm FinFETs using Monte Carlo Model and Drift-Diffusion Model with ballistic mobility. Version L-2016.12. Mountain View, CA, USA: Synopsys
19. Sentaurus Device User Guide, L-2016.12. Mountain View, CA, USA: Synopsys; 2016
20. BSIM-CMG. Multi-Gate MOSFET Compact Model (Technical Manual) [Online]. 2016. Available from: http://wwwdevice.eecs.berkeley.edu/bsim/?page=BSIMCMG
21. Datta S. Electronic Transport in Mesoscopic Systems. Cambridge, UK: Cambridge University Press; 1997
22. Sentaurus™ Device Monte Carlo. Sentaurus Band Structure User Guide, L-2016.12. Mountain View, CA, USA: Synopsys; 2016
23. Guo W, Choi M, Rouhi A, Moroz V, Eneman G, Mitard J, Witters L, et al. Impact of 3D integration on 7 nm high mobility channel devices operating in the ballistic regime. In: 2014 IEEE International Electron Devices Meeting (IEDM). IEEE; 2014. pp. 7-1
24. Moroz V, Smith L, Huang J, Choi M, Ma T, Liu J, Zhang Y, Lin XW, Kawa J, Saad Y. Modeling and optimization of group IV and III–V FinFETs and nano-wires. In: Proceedings of International Electron Devices Meeting; 2014. pp. 4-7
25. Moroz V, Huang J, Choi M, Smith L. Material engineering for 7nm FinFETs. In: ECS Transactions. Vol. 61, No. 3; 2014. pp. 103-110
26. Lenzi M, Palestri P, Gnani E, Reggiani S, Gnudi A, Esseni D, et al. Investigation of the transport properties of silicon nanowires using deterministic and Monte Carlo approaches to the solution of the Boltzmann transport equation. IEEE Transactions on Electron Devices. 2008;55(8):2086-2096
27. Guide, Virtuoso Spectre Circuit Simulator User. San Jose, CA, USA: Cadence; 2004
28. OpenCores. 2009. http://opencores.org/ (Cit. on pp. 63, 64)

7

Magnetoresistance and Structural Characterization of Electrospun La¹–Xsrxmno³ Nanowire Networks

1. Introduction

Colossal magnetoresistance (CMR) is a property of some materials, mostly manganese-based perovskite oxides that enables them to dramatically change their electrical resistance in the presence of a magnetic field, i.e., magnetoresistance (MR) [1, 2, 3]. To bring the CMR materials toward applications, it is still necessary to further optimize the sample processing to find the optimal microstructure, especially concerning grains in the nanometer range. The low-field magnetoresistance response (LFMR) of such manganese perovskite oxides like La1−xSrxMnO3 (LSMO) with different doping levels x is closely connected to the existence of interfaces and grain boundaries (GBs) within the samples [4, 5, 6, 7], and the high-field magnetoresistance (HFMR) was found to increase progressively on decreasing the grain size [8]. Therefore, the measurements of nanostructured or nano-sized samples are an important issue to provide a deeper understanding and further improvement of the MR effect, especially for the improvement of the behavior of devices based on the MR effect in reduced dimensions.

Commonly, the manganese perovskite samples studied in the literature are prepared as bulks or as thin films, mainly on SrTiO3 substrates. Nanometer-sized bridges are then prepared using lithography techniques or the focused ion-beam technique [9, 10]. The use of a substrate in the case of thin-film samples is always causing strain effects within the functional layer due to the lattice mismatch, which may play an important influence on the resulting magnetic properties [11]. The situation may be considerably different in nanostructures without substrate, being recently investigated in several types of nanoscale composites [12, 13, 14]. These nanostructures include nanorods, nanowires, nanotubes, and nanobelts; all of them having specific physical properties depending on the chosen preparation route.

In the present contribution, we have fabricated nanowire network fabrics of La1–xSrxMnO3 (LSMO) with different doping levels x by means of the electrospinning technique [15, 16, 17]. This technique is common for the fabrication of organic polymer nanostructures but can be modified by employing different precursors to deliver inorganic compounds. Up to now, only a small number of reports are dealing with magnetic nanostructures prepared in this way [17, 18, 19]. In the case of LSMO, exhibiting the CMR effect, the resulting nanowires are of polycrystalline nature with a high-aspect ratio (length up to 100 μm and diameters of about 230 nm) and show a large number of grain boundaries (GBs) within each individual nanowire. In Ref. [19], individual nanowires were separated from the as-spun networks and placed on a pre-patterned substrate. Electric contacts were prepared using Ti/Cu electrodes, and the nanowires could be measured individually, allowing to observe a dependence of the MR on the nanowire diameter. Single nanowires of CMR materials may be employed as sensitive gas sensor elements or electrodes [20, 21, 22].

However, the as-spun nanowires form a nonwoven fabric-like network, where numerous interconnects between the individual nanowires are formed in the final heat treatment step. As a result, the current flow through such a nanowire network fabric shows percolative character, and several sub-loops can be formed. The interconnects between the individual nanowires add additional crossover points for the currents and can enhance the tunneling transport across the interfaces, together with the GBs. This additional scattering of the electrons at the interfaces provided by the interconnects is lost when measuring only extracted parts of the nanowires as done in Ref. [19]. Furthermore, no information on the LSMO grain size of their nanowires was presented. An analysis of the grain sizes within our nanowires showed values ranging between 10 and 32 nm. Therefore, it is obvious that the LSMO grains within the present nanowires are smaller as compared to, e.g., Ref. [8].

Therefore, we may expect interesting new properties of this new class of magnetic material. Furthermore, the nanowire network fabrics are an extremely lightweight material with a density of about 0.084 g/cm3, which is considerably less than the theoretical density of 6.5 g/cm3 [23]. Furthermore, there is no sample size limitation imposed by the fabrication technique, as electrospinning may produce very large sample sizes [24]. This makes such fabrics interesting for applications in bulk form, whenever the weight of the sample counts.

In order to achieve a better understanding of the transport properties through these nanowire network fabrics, we also performed a thorough microstructure analysis including scanning electron microscopy (SEM) and transmission electron microscopy (TEM).

2. Experimental procedures

The electrospinning precursor is prepared by dissolving La, Sr, and Mn acetates in PVA (high-molecular-weight polyvinyl alcohol). The PVA is slowly added to the acetate solution with a mass ratio of 2.5:1.5. This solution is stirred at 80°C for 2 h and then spun into cohering nanofibers by electrospinning. To remove the organic compounds and to form the desired LSMO phase, the sample is subsequently heat treated in a lab furnace. An additional oxygenation process is required to obtain the correct phase composition. The constituent phase was checked by means of X-ray diffraction (XRD) and EDX analysis. Further details about the electrospinning process of ferromagnetic and superconducting nanowires are given elsewhere [25, 26, 27, 28].

Figure 1(a, b) presents photographs of an as-grown La0.8Sr0.2MnO3 nanowire fabric. The nanowire fabric consists of polymer nanowires containing the ceramic precursor material. The as-grown fabric has a white color, and the entire fabric sample is fully flexible. Figure 1(c, d) finally presents the fully reacted sample after having received the full thermal treatment. The reacted sample shows a fully black color, indicating the completed chemical reaction. As a result, the final nanowire network fabric is extremely thin and brittle. Here, it is important to note that the sample size shrunk to about one sixth of its original size. This shrinkage has to be considered for the application of such fabric-like materials. In the thermal treatment, numerous interconnects between the individual nanowires are formed, which are essential for the resulting current flow through the sample.

Figure 1. Images of the electrospun samples fabricated from the La0.8Sr0.2MnO3 precursor. Images (a) and (b) present a view of an as-prepared sample before thermal treatment, whereas images (c) and (d) give a La0.8Sr0.2MnO3 sample after the whole annealing process applied. As can be directly seen from the images, the size (area) of the sample shrinks to one sixth as compared to the original one after the thermal treatment.

The entire nanowire network was electrically connected by means of silver paint and Cu wires (50 μm diameter) to the sample holder. Due to the high fragility of the ceramic sheet, a pseudo four-point configuration is realized where the current and voltage links connect immediately on the sample contacts. This arrangement is presented in Figure 2. The magnetoresistance is measured in a 10/12 T bath cryostat (Oxford Instruments Teslatron) with a Keithley source meter (model 2400) as a current source, and the voltage is recorded using a Keithley 2001 voltmeter.

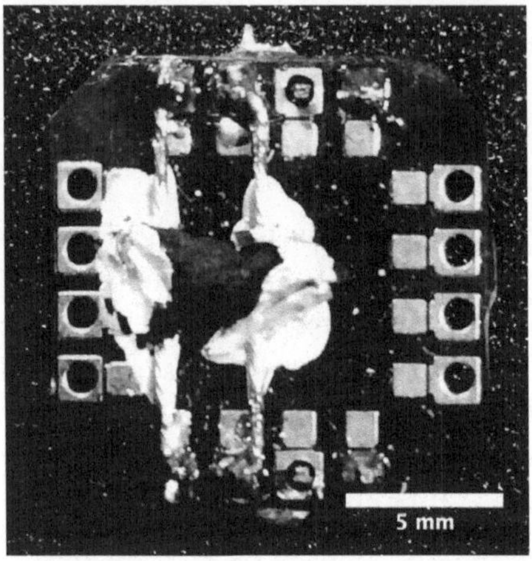

Figure 2. Nanowire network sample with electrical contacts for the quasi four-point measurement.

The constituent phases of the samples were determined by means of a high-resolution automated RINT2200 X-ray powder diffractometer using Cu-Kα radiation (40 kV, 40 mA) Figure 3. SEM imaging was performed using a Hitachi S800 scanning electron microscope operating at a voltage of 10 kV, and the TEM analysis was performed by a JEOL JSM-7000F transmission electron microscope (200 kV, LaB6 cathode). For TEM imaging, pieces of the nanowire network fabrics were deposited on carbon-coated TEM grids. High-resolution TEM and EBSD were performed on selected nanowire sections being thin enough for electron transmission (Figure 6).

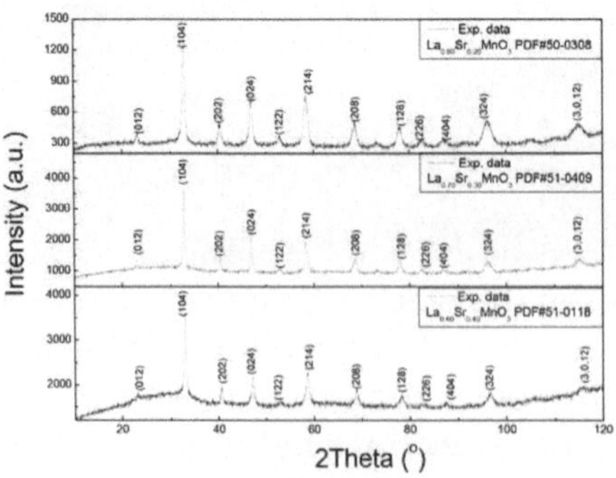

Figure 3. XRD measurements on all three types of LSMO nanowire fabric samples.

The magnetization of the nanowire networks was measured using a SQUID magnetometer (Quantum Design MPMS3) with ±7 T magnetic field applied perpendicular to the sample surface, using a piece of the nanowire network fabric with a size of 14.86 mm2.

3. Results and discussion

3.1. Microstructure

Scanning electron microscopy revealed an average diameter of the resulting nanowires of around 220 nm and a length of more than 100 µm. Fabric-like nanowire networks with numerous interconnects are formed after the heat treatment. The individual nanowires are polycrystalline with a grain size of about 10–30 nm, which corresponds to the dimensions obtained via transmission electron microscopy and electron backscatter diffraction (EBSD) analysis.

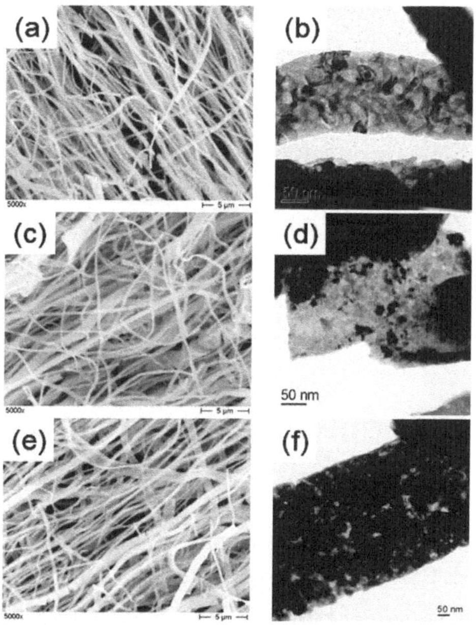

Figure 4. SEM images of the nanowire network fabrics at 5000× magnification (first column) and at higher magnification (10,000×, second column) for all LSMO samples studied here. Images (a) and (b) show the sample x = 0.2, (c) and (d) the sample x = 0.3, and (e) and (f) the sample x = 0.4. The low-magnification images (a, c, and e) present the individual nanowires forming the network fabrics and the numerous interconnects between them. The images with higher resolution (b, d, and f) reveal the polycrystalline structure of the LSMO nanowires. Several small LSMO grains form stacks building up a nanowire.

This is presented in Figure 4 giving SEM images of the nanowire network fabrics at 5000× magnification (first column) and at higher magnification (10,000×, second column) for all LSMO samples studied here. Figure 4(a) and (b) shows the sample x = 0.2, (c) and (d) the sample x = 0.3, and (e) and (f) the sample x = 0.4. The low-magnification images (a, c, and e) present the individual nanowires forming the network fabrics and the numerous interconnects between them. These interconnects, formed during the final heat treatment step, are essential for the current flow through the entire sample perimeter and provide additional scattering at their interfaces. The images with higher resolution (b, d, and f) reveal the polycrystalline structure of the LSMO nanowires. Several small LSMO grains form stacks building up a nanowire. Furthermore, it is obvious from these images that the LSMO grains do not show a texture but are randomly oriented with high-angle grain boundaries between them.

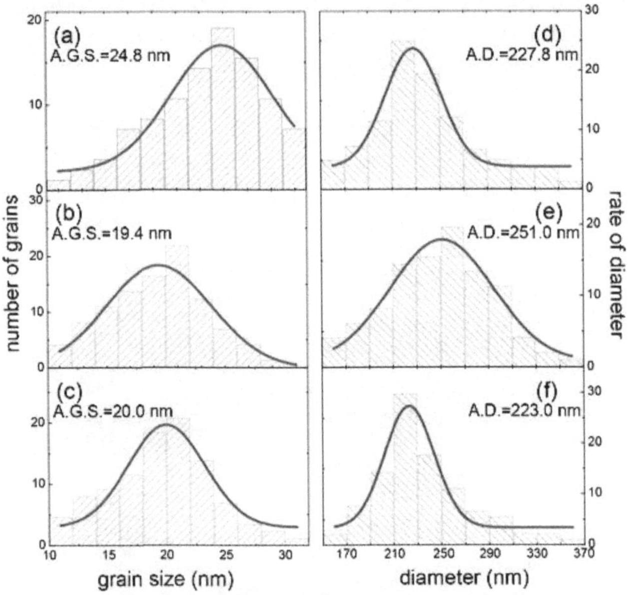

Figure 5. Statistics of the nanowire average grain size (AGS) and the diameter (AD): (a) and (d) La0.8Sr0.2MnO3, (b) and (e) La0.7Sr0.3MnO3, and (c) and (f) La0.6Sr0.4MnO3.

Figure 5 presents the detailed analysis of the nanowire diameters and the LSMO grain size determined from several SEM and TEM images. Graphs (a), (b), and (c) show the grain size analysis. The average values were determined using a Gauss fit to the data (indicated by a red line). For the sample x = 0.2, the grain size determined from TEM images varies between 10 and 32 nm with an average of 24.8 nm. Sample x = 0.3 shows grain sizes ranging between 10 and 30 nm with an average of 19.4 nm, and sample x = 0.4 shows an equal distribution of grain sizes with an average grain size of 20 nm. For all samples, the grain sizes determined from the TEM images are larger than the average grain size estimated from XRD, indicating that there are multiply connected domains in the sample. On the sample x = 0.2, we also performed an EBSD analysis of the grain orientations [29], demonstrating the presence of a large number of high-angle grain boundaries within an individual nanowire. The graphs of the nanowire diameters (d, e, f) reveal an average diameter of 227.8 nm for the sample x = 0.2, 251.0 nm for sample x = 0.3, and 223.0 nm for sample x = 0.4. Note that also here the sample x = 0.4 exhibits a much narrower size spectrum as compared to the other samples studied here.

Figure 6 shows finally some high-resolution TEM images of all three LSMO samples studied here. Figure 6(a) and (b) gives grains and their grain boundaries of sample x = 0.2, (c) and (d) images with the same magnification of sample x = 0.3, and (e) and (f) those of sample x = 0.4. For all samples, the LSMO grain size is around 20 nm, and the grains are randomly oriented, resulting in the presence of high-angle grain boundaries between the grains. Furthermore, some grain clusters connected with low-angle grain boundaries are also visible, which explains the differences between the grain sizes determined from X-ray data and from TEM images and EBSD [29]. The nanowires are built up from stacks of several individual LSMO grains.

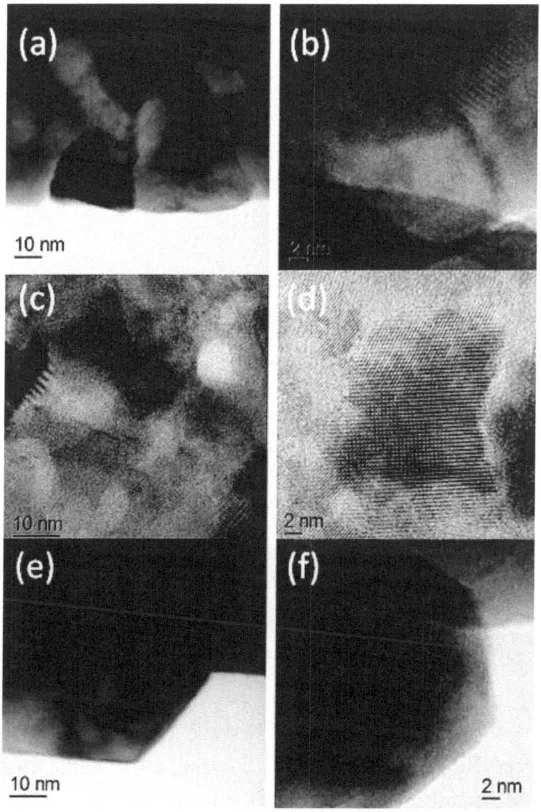

Figure 6. High-resolution TEM images with two magnifications of all three types of samples studied here. (a, b) sample x = 0.2, (c, d) sample x = 0.3, and (e, f) sample x = 0.4. The LSMO grains of all samples have sizes of ~20 nm, and the grains are randomly oriented with high-angle grain boundaries between them.

3.2. Magnetization data

Figure 7 presents the magnetization data obtained for all three compositions. The soft magnetic character of the LSMO fabric samples is clearly revealed. From the Sr-doping level x = 0.2 to x = 0.4, the saturation magnetization M s is at 10 K 52.12, 58.13, 55.03 emu/g, and at 300 K, and the results are 18.10, 33.09, and 24.62 emu/g, respectively. These values are lower than the ones from the corresponding bulk materials (56 emu/g [30], 33.09 emu/g [31], 24.62 emu/g [32]). The x = 0.3 sample shows the highest M s among the three compositions, while the x = 0.2 sample shows the lowest ones at both 10 and 300 K. This matches the conclusion drawn from bulk materials that the optimized doping level around x = 0.33 supplies the strongest magnetism. The M s value of all our samples being lower than the bulk values can be attributed to the smaller LSMO grain size of our nanowire network fabric samples.

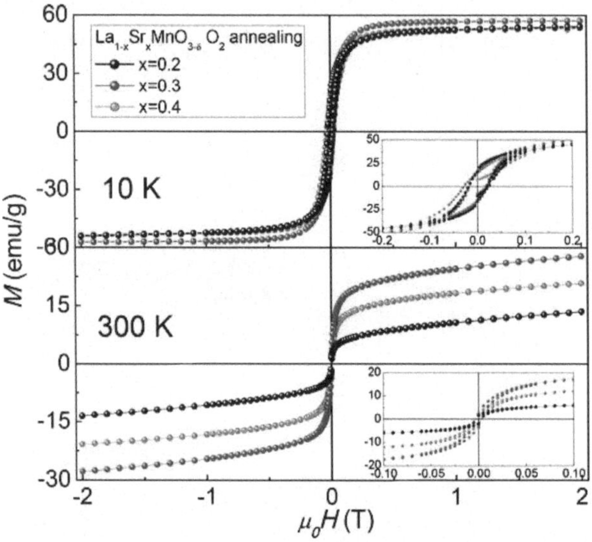

Figure 7. M(H) measurement of the LSMO nanowires at T = 10 and 300 K for the LSMO nanowire network fabrics with x = 0.2, 0.3, and 0.4. The insets present the low-field part of the M(H) loops.

3.3. MR data

With respect to the literature, the MR data can be divided into two different regimes. According to Ref. [8], the high-field magnetoresistance (HFMR) behavior sheds light on the influence of the sample microstructure via the interface response. Therefore, analyses of the electronic transportation properties and of the magnetoresistive effects of the nanowire networks were carried out by four-probe measurements in external magnetic fields up to 10 T. Firstly, we have a look at the high-field regime. Figure 8 presents the resistance measurements for the samples x = 0.2, 0.3, and 0.4 with the temperature swept from 5 to 300 K. The magnetic field is applied perpendicular to the surface of the nanowire network, i.e., the current flow is mainly perpendicular to the applied field. Magnetoresistance plots are calculated from the data obtained by these sweep measurements at different field strengths using the relation

$$MR[\%] = \frac{R_H(T) - R_0(T)}{R_0(T)}. \quad (1)$$

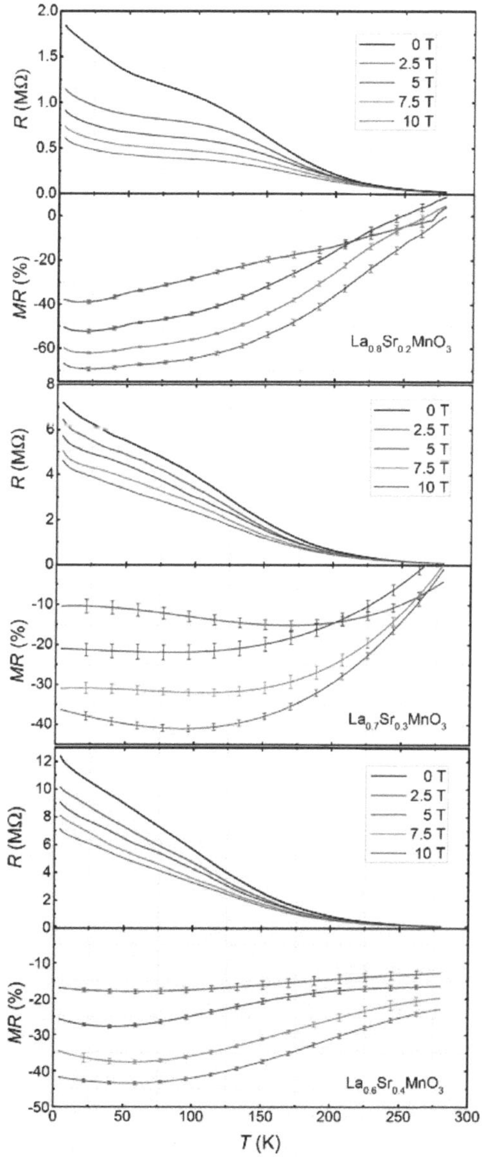

Figure 8. Resistance and MR ratio of all LSMO samples measured up to 10 T applied magnetic field (H ⊥ sample surface) in the range 2 K < T < 275 K: (a) x = 0.2, (b) x = 0.3, and (c) x = 0.4.

As result, we find a maximum MR for the sample x = 0.2 at low temperatures and 10 T applied magnetic field of 70%.

Table 1 summarizes our findings at three selected magnetic fields for the three doping levels studied here, and the last row of Table 1 gives the maximum MR obtained, together with the respective temperature.

Table 1. MR comparison of all LSMO samples measured here at 1, 2.5, and 10 T applied magnetic field.

Doping level	**MR(%) at 1 T**		
x	100 K	270 K	MR max.
0.2	10.24	130	20.66 (22 K)
0.3	6.23	5.98	14.54 (203 K)
0.4	23.58	7.69	26.01 (63 K)
	MR(%) at 2.5 T		
0.2	27.69	1.81	39 (14 K)
0.3	13.2	6.32	15.18 (169 K)
0.4	17.48	13.00	17.97 (57 K)
	MR(%) at 10 T		
0.2	64.12	4.62	69.28 (25 K)
0.3	40.88	5.84	40.93 (92 K)
0.4	41.99	23.47	43.45 (54 K)

The last row gives the maximum MR obtained, together with the respective temperature.

At 100 K, sample x = 0.4 shows the highest MR of 23.58% (close to the overall maximum of 26%) at 1 T applied field. In higher applied fields, the maximum value of the MR shifts to sample x = 0.2. A maximum MR ratio of 26.01 at an applied field of 1 T (63 K) is obtained for the sample x = 0.4. At 2.5 T applied field, the maximum MR of 17.97 (57 K) is found in sample x = 0.4.

At 10 T applied magnetic field, the maximum MR rate of 69.28% (25 K) is found in sample x = 0.2. Note also that the highest MR of sample x = 0.3 is obtained at a relatively high temperature, ranging from 203 K (1 T) to 92 K (10 T). Secondly, Figure 9 shows the MR ratio of sample x = 0.2 as a function of applied field for various temperatures. The MR is found to increase on lowering the temperature down to 50 K, whereas the MR at low temperature (3 K) is found to be smaller as compared to the 50 K data. The inset of Figure 9 presents the LFMR at an applied magnetic field of 0.1 T for the sample x = 0.2, showing a maximum LFMR of 5.2% at T = 50 K.

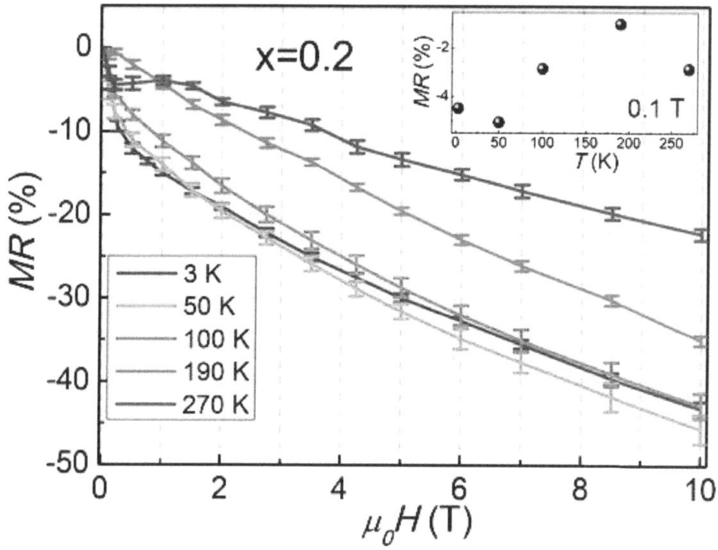

Figure 9. Magnetoresistance of La0.8Sr0.2MnO3 nanowires as function of the applied field for various temperatures. The inset gives the LFMR at 0.1 T applied magnetic field.

3.4. Discussion

From the graphs and the tables presented here, three main features can be deduced: Firstly, there is a suppression of the metal-insulator transition. This provides another evidence of the size effect. Nanoscale grains are always accompanied by a large number of grain boundaries, which enhance electron scattering. The metallic behavior is negatively influenced by the Coulomb blockade [13], and as a result, an upturn of the resistivity appears at low temperatures. On the other hand, the influence from certain size effects varies with the doping level. So, a step-shape resistance behavior can be observed for the x = 0.2 sample, which indicates that the metal-insulator transition is not completely suppressed. However, in the x = 0.4 sample, the resistance decreases almost linearly from 5 K to 200 K. Secondly, the x = 0.3 sample with the highest saturation magnetization shows the weakest MR effect, and inversely, the x = 0.2 sample exhibits the strongest MR effect at low temperatures. At 2.5 T, the maximum of the MR is already over 40%, which points to an obvious LFMR effect. The increase of the MR with reducing magnetization has been reported by Balcells et al. [8]. In their work, the variation of the grain size is the main reason of the behavior observed. In contrast to this, there is no obvious grain size difference between our samples. The dependence of the MR on the magnetization requires further investigation. Thirdly, at around 200 K with the doping level x = 0.2 and 0.3, the MR curves at 2.5 T have a crossover with the curves at other fields, which means that above 200 K, the absolute value of the magnetoresistance does not monotonously increase with the external field.

To enable a comparison with data on other LSMO sample types but with the same chemical composition (x = 0.2) published in the literature, we summarize some data at 1 T applied magnetic field in Table 2 and compare them with our results.

Table 2. Magnetoresistance (MR) comparison of different sample types (nanowires, nanoparticles, and powder) of La0.8Sr0.2MnO3 samples studied in the literature at 1 T applied magnetic field.

$La_{0.8}Sr_{0.2}MnO_3$	MR(%) at 1 T		
Type	100 K	270 K	MR max
Nanowires [33]	11.2	3.93	20.66 (22 K)
Nanoparticles [34]	19.36	4.66	18.7 (79 K)
Powder [35]	13.56	6.47	17.66 (50 K)

From Table 2, we see that the nanowire sample x = 0.2 exhibits a larger MR of 20.66% as compared to the nanoparticles and the powder samples of the same chemical composition studied in the literature, however, at some lower temperature. This points clearly to the important contribution of the much smaller LSMO grains in our nanowire network fabric samples, and the additional scattering at the numerous interconnects provides extra MR.

Jugdersuren et al. [19] reported a large LFMR at room temperature in their LSMO nanowires extracted from the networks produced by electrospinning and showed a dependence of the LFMR on the nanowire diameter, but no information on the LSMO grain size was presented. However, the LFMR at 300 K for our sample is comparable to their data, even though the chemical composition is somewhat different. Nevertheless, this demonstrates that the LFMR as well as the HFMR can be considerably enhanced by reducing the nanowire diameter as well as the LSMO grain size.

4. Conclusion

Nonwoven nanowire networks of LSMO with three doping levels x = 0.2, 0.3, and 0.4 were fabricated by means of electrospinning. The magnetoresistance effect of these nanowire fabrics was measured up to 10 T applied field between 5 K and 300 K. The reduced dimensions of the nanowires and the large number of interconnects between them were found to increase the MR effect as compared to bulk samples. An MR ratio of 70% was observed for the sample x = 0.2 at low temperatures and 10 T applied magnetic field.

References

1. Jin S, Tiefel TH, McCormack M, Fastnacht RA, Ramesh R, Chen LH. Thousandfold change in resistivity in magnetoresistive La-Ca-Mn-O films. Science. 1994;264:413-415

2. Rodriguez LM, Attfield JP. Cation disorder and size effects in magnetoresistive manganese oxide perovskites. Physical Review B. 1996;54:R15622-R15625

3. Ramirez AP. Colossal magnetoresistance. Journal of Physics: Condensed Matter. 1997;9:8171-8199

4. Hwang HY, Cheong S-W, Ong NP, Batlogg P. Spin-polarized intergrain tunneling in La2/3Sr1/3MnO3. Physical Review Letters. 1996;77:2041

5. Mahesh R, Mahendiran R, Raychaudhuri AK, Rao CNR. Effect of particle size on the giant magnetoresistance of La0.7Ca0.3MnO3. Applied Physics Letters. 1996;68:2291

6. Li XW, Gupta A, Xiao G, Gong GQ. Low-field magnetoresistive properties of polycrystalline and epitaxial perovskite manganite films. Applied Physics Letters. 1997;71:1124

7. Kar S, Sarkar J, Ghosh B, Raychaudhuri AK. Effect of grain boundaries on the local electronic transport in nanostructured films of colossal magnetoresistive manganites. The Journal of Nanoscience and Nanotechnology. 2007;7:2051

8. Balcells L, Fontcuberta J, Martinez B, Obradors X. High field magnetoresistance at interfaces in manganese perovskites. Physical Review B. 1998;58:R14697

9. Liu D, Liu W. Room temperature ultrahigh magnetoresistance nanostructure (La2/3Sr1/3)MnO3 films growth on SrTiO3 substrate. Ceramics International. 2012;38:2579

10. Marin L, Morelloì L, Algarabel PA, Rodiguez LA, Magen C, De Teresa JM, et al. Enhanced magnetotransport in nanopatterned manganite nanowires. Nano Letters. 2012;11:103

11. Hwang HY, Iwasa Y, Kawasaki M, Keimer B, Nagaosa N, Tokura Y. Emergent phenomena at oxide interfaces. Nature Materials. 2014;14:423

12. Li L, Liang L, Wu H, Zhu X. One-dimensional perovskite manganite oxide nanostructures: Recent developments in synthesis, characterization, transport properties, and applications. Nanoscale Research Letters. 2016;11:121

13. Sarkar T, Kamalakar MV, Raychaudhuri AK. Electrical transport properties of nanostructured ferromagnetic perovskite oxides La0.67Ca0.33MnO3 and La0.5Sr0.5CoO3 at low temperatures (5 K 0.3 K) and high magnetic field. New Journal of Physics. 2012;14:033026

14. Dwivedi GP, Kumar M, Shahi P, Barman A, Chatterjee S, Ghosh AK. Low temperature magnetic and transport properties of LSMO-PZT nanocomposites. RSC Advances. 2015;5:30748

15. Li D, McCann JT, Xia YN. Electrospinning: A simple and versatile technique for producing ceramic nanofibers and nanotubes. Journal of the American Ceramic Society. 2006;89:1861

16. Wu H, Pan W, Lin D, Li H. Electrospinning of ceramic nanofibers: Fabrication, assembly and applications. Journal of Advanced Ceramics. 2012;1:2

17. Li D, Herricks T, Xia YN. Magnetic nanofibers of nickel ferrite prepared by electrospinning. Applied Physics Letters. 2003;83:4586
18. Yensano R, Pinitsoontorn S, Amornkitbamrung V, Maensiri S. Fabrication and magnetic properties of electrospun La0.7Sr0.3MnO3 nanostructures. Journal of Superconductivity and Novel Magnetism. 2014;27:1553
19. Jugdersuren B, Kang S, DiPietro RS, Heiman D, McKeown D, Pegg IL, et al. Large low field magnetoresistance in La0.67Sr0.33MnO3 nanowire devices. Journal of Applied Physics. 2011;109:016109
20. Liu Y, Sun X, Li B, Lei Y. Tunable p-n transition behaviour of a p-La0.67Sr0.33MnO3/n-CeO2 nanofibers heterojunction for the development of selective high temperature propane sensors. Journal of Materials Chemistry A. 2014;2:11651
21. Xu D, Luo L, Ding Y, Xu P. Sensitive electrochemical detection of glucose based on electrospun La0.88Sr0.12MnO3 nanofibers modified electrode. Analytical Biochemistry. 2015;489:38
22. Zhi M, Koneru A, Yang F, Manivannan A, Li J, Wu N. Electrospun La0.8Sr0.2MnO3 nanofibers for a high-temperature electrochemical carbon monoxide sensor. Nanotechnology. 2012;23:305501
23. Armstrong TJ, Virkar AV. Performance of solid oxide fuel cells with LSGM-LSM composite cathodes. Journal of the Electrochemical Society. 2002;149(12):A1565-A1571
24. Huang ZM, Zhang YZ, Kotaki M, Ramakrishna S. A review on polymer nanofibers by electrospinning and their application in nanocomposites. Composites Science and Technology. 2003;63:2223-2253
25. Zeng XL, Koblischka MR, Hartmann U. Synthesis and characterization of electrospun superconducting (La,Sr)CuO4 nanowires and nanoribbons. Materials Research Express. 2015;2:095022
26. Koblischka MR, Zeng XL, Karwoth T, Hauet T, Hartmann U. Transport and magnetic measurements on Bi-2212 nanowire networks prepared via electrospinning. IEEE Transactions on Applied Superconductivity. 2016;26:1800605
27. Zeng XL, Koblischka MR, Karwoth T, Hauet T, Hartmann U. Preparation of granular Bi-2212 nanowires by electrospinning. Superconductor Science and Technology. 2017;30:035014
28. Koblischka MR, Zeng XL, Laurent F, Karwoth T, Koblischka-Veneva A, Hartmann U, et al. Characterization of electrospun BSCCO nanowires with reduced preparation temperature. IEEE Transactions on Applied Superconductivity. 2018;28:7200505
29. Koblischka-Veneva A, Koblischka MR, Zeng XL, Schmauch J. Journal of Magnetism and Magnetic Materials. submitted for publication
30. Dominiczak M, Ruyter A, Limelette P, Laffez IM, Giovannelli F, Rossell MD, et al. Effects of nanocracks on the magnetic and electrical properties of La0.8Sr0.2MnO3 single crystals. Solid State Communications. 2009;149:1543

31. Huang BX, Liu YH, Yuan XB, Wang CJ, Zhang RZ, Mei LM. The unusual magnetotransport properties of La0.67Sr0.33MnO3 with Nb2O5 addition. Journal of Magnetism and Magnetic Materials. 2004;280:176

32. Vijayanandhini K, Kutty TRN. Magnetically tuneable nonlinear electronic properties of ZnO/La(Sr)MnO3 composites. Journal of Physics D: Applied Physics. 2006;39:2902

33. Karwoth T, Zeng XL, Koblischka MR, Hartmann U, Chang C, Hauet T, et al. Magnetoresistance and structural characterization of electrospun La1-xSrxMnO3 nanowire networks. Solid State Communications. submitted for publication

34. Nagabhuhana BM, Sreekanth Chakradhar RP, Ramesh KP, Prasad V, Shivakumara C, Chandrappa GT. Magnetoresistive studies on nanocrystalline La0.8Sr0.2MnO3 manganite. Physica B. 2008;403:3360

35. Grossin D, Noudem JG. Synthesis of fine La0.8Sr0.2MnO3 powder by different ways. Solid State Sciences. 2004;6:939

Index

A

Acetate 57
Active 74, 96, 98, 101
Alkaline 50
Alloys 67
Al-masked 17
Alumina 19
Aluminum 19
Ammonia 51
Analysis 26-27, 42, 44, 75, 84, 110, 114
Anisotropy 2-3, 11
Autoclave 38

B

Battery 37-38, 41, 46-47
B-doped 19
Bulletin 66

C

Capacity 41
Capillary 20
Carbon 4, 11-13, 16, 21, 34-35
Catalysts 74
Cathode 20, 37-39, 41, 46-48
Ceramics 121
Clusters 52, 62
Components 106
Copper 20
Crystals 22, 40

D

Density 17, 24, 37, 50, 54, 58, 64-66, 110
Deposition 3-4, 18, 32-33, 57, 72, 89
Design 104
Detectors 15
Diameters 53, 98
Dipole 11
Doping 119
Droplets 72

E

Electrodes 1, 44, 110
Electrons 2, 77
Emission 15, 26, 30-31, 35, 73, 77-80, 82-83, 86, 88
Energy 21, 37, 49, 57, 61, 73, 80-82, 85-86, 101-105
Equimolar 55, 61
Erosion 60
Evolution 88

F

Fabrics 110, 113
Ferrocene 4
Fusion 22

G

Geometry 79
Glucose 27
Grains 114-115

Graphite 21, 23

Growth 3, 16-21, 32, 50-52, 54-57, 59-66, 68-69, 71-72, 74, 76, 82, 87-88

H

Hexagonal 50, 60-61, 64

Hybrid 29

Hydrogen 20-21

Hysteresis 8, 11

I

Industry 71

Insulator 1-2

Intensity 73, 77, 85

Interior 4

Inversion 93

Isotropic 43

K

Kinetic 37

L

Liquid 87

Lithium 38, 43-48

M

Magnetism 122

Manganese 43-44, 47-48, 109, 121

Manganite 121

Metastable 60

Methane 21

Microscope 112

Microscopy 28, 39, 41, 110, 113

Microwave 19

Mixture 50

Mobility 108

Modeling 1, 106

Molecules 26-27

Morphology 39, 46, 50, 52, 55, 63, 65, 68

N

Nanofibers 121-122

Nanograss 35

Nanorods 22, 47, 52-53, 55-56, 58-62, 64, 66-68

Nanoscale 36, 71, 76

Nanowire 3, 11, 28-30, 34, 36-39, 45, 50-52, 55, 57, 59-60, 63, 65-66, 68, 72, 87, 89, 97, 108, 110-116, 120, 122

Nature 12, 30-31

Nitrate 59

Nitrogen 33

Nutrient 61-62, 64-65

O

Oscillator 95, 99-100

Oxidation 43

Oxygen 17

P

P-channel 92, 94

Photonics 71, 87

Plasma 16-18, 20-21, 31, 33

Precursors 59

Q

Quantum 1, 73, 76-77, 80, 86, 89

R

Resistance 15, 119

Resonance 39, 43

S

Saturation 8

Seedless 59, 66, 68

Sensing 36, 67

Sensitive 36

Sensors 27, 67

Signal 6

Silicon 28, 71, 87-89, 91, 108

Spectra 39, 42, 44, 77-78, 82-83
Spectrums 61
Stress 94
Suspension 18
Symposium 107
Synopsys 108
Systems 37, 86

T

Transistor 91-92

U

Unique 29

V

Variation 101
Velocity 95
Vibration 42

X

Xylene 4

9 789360 848729

Printed at Repro India Ltd.